TEACHING MACHINES

TEACHING MACHINES

THE HISTORY OF PERSONALIZED LEARNING

AUDREY WATTERS

THE MIT PRESS   CAMBRIDGE, MASSACHUSETTS   LONDON, ENGLAND

© 2021 Audrey Watters

All rights reserved. No part of this book may be reproduced in any form by any electronic or mechanical means (including photocopying, recording, or information storage and retrieval) without permission in writing from the publisher.

We have made every effort to trace the rights holder for the cover image. Rights holders may contact us at https://mitpress.mit.edu/contact-us and full credit will be given in subsequent printings.

This book was set in ITC Stone and Avenir by New Best-set Typesetters Ltd. Printed and bound in the United States of America.

Library of Congress Cataloging-in-Publication Data

Names: Watters, Audrey, author.
Title: Teaching machines : the history of personalized learning / Audrey Watters.
Description: Cambridge, Massachusetts : The MIT Press, [2021] | Includes bibliographical references and index.
Identifiers: LCCN 2020024477 | ISBN 9780262045698 (hardcover)
Subjects: LCSH: Educational technology—History—20th century. | Programmed instruction—History—20th century. | Skinner, B. F. (Burrhus Frederic), 1904-1990 | Web-based instruction—History—20th century.
Classification: LCC LB1028.3 .W383 2021 | DDC 371.33—dc23
LC record available at https://lccn.loc.gov/2020024477

10 9 8 7 6 5 4 3 2 1

For teachers . . . for *human* teachers, not teaching machines

If by a miracle of mechanical ingenuity, a book could be so arranged that only to him who had done what was directed on page one would page two become visible, and so on, much that now requires personal instruction could be managed by print.

—Edward Thorndike, *Education, a First Book* (1912)

CONTENTS

INTRODUCTION 1

1 B. F. SKINNER BUILDS A TEACHING MACHINE 19

2 SIDNEY PRESSEY AND THE AUTOMATIC TEACHER 35

3 "MECHANICAL EDUCATION WANTED" 61

4 THE COMMERCIALIZATION OF B. F. SKINNER'S FIRST MACHINES 81

5 B. F. SKINNER TRIES AGAIN 107

6 PROGRAMMED INSTRUCTION: IN THEORY AND PRACTICE 135

7 IMAGINING THE MECHANIZATION OF TEACHERS' WORK 149

8 HOLLINS COLLEGE AND "THE ROANOKE EXPERIMENT" 167

9 TEACHING MACHINES INC. 179

10 B. F. SKINNER'S DISILLUSIONMENT 195

11 **PROGRAMMED INSTRUCTION AND THE PRACTICE OF FREEDOM** 213

12 **AGAINST B. F. SKINNER** 231

CONCLUSION 245

ACKNOWLEDGMENTS 265
NOTES 269
INDEX 301

INTRODUCTION

In 2012, Sal Khan sat down with *Forbes* journalist Michael Noer and recorded an eleven-minute video on "The History of Education" as part of the media tour Khan was undertaking to promote his new book, *The One World School House*. Although he had no formal teacher training, Khan, an MIT graduate and a former hedge fund analyst, had become something of an education celebrity over the course of the previous few years, initially setting out to tutor his cousin in math but eventually expanding his audience by posting a series of short, explanatory videos about all kinds of topics to YouTube—videos that received millions and millions of views. These videos had a decidedly low-budget look and feel—made with a Wacom tablet, using video capture to record Khan's screen and voice but never his face or body, as he casually talked through various concepts, drawing graphs and figures with a stylus. Khan's videos attracted the attention of Silicon Valley investors and, most notably perhaps, of billionaire education philanthropist Bill Gates, who introduced Khan at the 2011 TED Talks. There, Khan gave a presentation, "Let's use video to reinvent education," which has since been watched online almost five million times.[1]

Khan's online video-based instruction was almost universally lauded by the press and by pundits. He was described as the savior of education. *Bloomberg Businessweek* dubbed him "The Messiah of Math." *Time* wondered if Khan was "The New Andrew Carnegie." *Slate* claimed that "his folksy lectures are revolutionizing how kids learn math and science." *Wired Magazine* argued that the nonprofit Khan had formed, Khan Academy, was "changing the rules of education." *Fast Company*, which described him as "Bill Gates' favorite teacher," cheered Khan Academy's potential to "disrupt education."[2]

This "disruption," as these publications and philanthropists saw it, was meant to upend a public school system that stifles creative inquiry and independent thought; that emphasizes standardization over personalization; that, as Khan claimed in his book (and in the *Forbes* video), demands students be separated by age into "buckets" and move to the sound of bells "in lockstep."[3] It is a system, they contended, that was established almost two hundred years ago and has not changed since. This "industrial model of education," sometimes derided as the "Prussian model of education," is inflexible and outmoded and, as such, requires sweeping technological upgrades.

Khan's "History of Education" video credits Horace Mann with importing this model to the United States in 1840. Just thirty years later, Khan explains, "you get to a situation where public education is actually fairly commonplace," although, admittedly, hardly standardized in any way. To address geographic discrepancies, Khan narrates, "in 1892 . . . you had a Committee of Ten." "Sounds somewhat Orwellian," the

Forbes journalist quips. "Somewhat Orwellian," Khan echoes in agreement, "but it is literally ten gentlemen, led by the President of Harvard, to determine what should happen" in schools. "It was actually this Committee of Ten that decided there should be twelve years of compulsory education," Khan explains, and that decreed what subjects students must take in their junior or senior year of high school—indeed, "that you should even have this notion of high school" at all.

"So this is standardization," Noer adds, summarizing Khan's points. "We have now a public standardized education. . . . It was forward thinking for 120 years ago," he says, "but what is interesting here is that we basically get stuck there." Khan agrees. Education, he says, "is static to the present day."

Until the computer, the internet, and Khan Academy came along, that is. Khan and Noer's history—sketched on a timeline Khan calls "the sweep of time"—skips from the the 1890s to the 1990s. Now, says Khan, it is no longer necessary "to batch everyone into these buckets. Everyone can go at their own pace, get feedback at their own pace. . . . Class time can be liberated."

For the very first time, technology will "make the classroom more human, more interactive" and free from the "Prussian model" of schooling.[4] For the first time in the history of education, as Khan and Noer tell it, learning will be personalized.

There's at least one problem with the way Khan tells it: the history is all wrong.

When the "industrial model" gets invoked in discussions about educational politics and policies, the phrase—

intentionally or not—glosses over and distorts history. Prussia was hardly industrialized in 1763 when Frederick the Great founded the country's public education system. Prussia lagged behind other parts of Europe—certainly behind Britain, which did *not* have a public school system at the time—in industrializing its economy.

One of the key features of the Prussian education system was—as Khan notes correctly in his *Forbes* video—the collection of taxes to pay for free schooling. But arguably the most important feature—and the one that education reformer Horace Mann was keen on when he visited Prussia in 1843—was that teachers should be *professionals*, trained in specialized colleges. It is striking that this part—the professionalization of educators—gets erased from the story Khan relates.

Granted, some of that erasure comes with the pace at which Khan moves through his timeline. It's a ten-minute video, after all. Much of his narration is sloppily worded, as part of the schtick of Khan Academy is that its videos appear off-the-cuff and casual. And while tightly scripting a lesson like it's a stage production has plenty of drawbacks, how one presents and explores ideas is worthy of care and intentionality, particularly if you're building an intellectual framework for someone else to navigate and rely upon. (The same goes for writers, no doubt.) Words and phrasing matter. The Committee of Ten did not *invent* the institution of high school, as the "history of education" video implies. The Committee did not *decide* anything; not really. They *recommended* twelve years of compulsory education. Contrary to the inference, there was no semblance of a federal system or federal requirements for education in 1892 or for many more years to come. There are still no federal standards for curriculum.

Even if you forgive Sal Khan for misspeaking or poorly articulating his point (and I'm not sure we should if he is, indeed, the "messiah" of education), many of his claims and inferences are inaccurate. At best, his story is missing a lot of context. In 1892, only a tiny fraction of American teenagers attended, let alone graduated from high school. What high schools existed at the time were mostly located in large urban areas in the North, where they served academically talented students—almost exclusively white students whose parents could afford for them to not work. An even smaller number of students actually went on to college. Nonetheless the Committee of Ten felt that those students who did attend high school should have a rigorous academic training to prepare them for their careers, sure, but also to help them become intelligent, informed citizens.

Much has changed—not only in high school demographics—since 1892, and yet those who like to repeat this tale of an "industrial model" often insist, as Khan does in the *Forbes* video, that the school system has been "static to the present day."

To call the US education system "static" from 1892 onward is to narrate a history of education that is woefully inaccurate—offensively so, in fact. It is to ignore, for example, the Supreme Court decision just a few years later in 1896, *Plessy v. Ferguson*, that upheld a racially segregated school system as "separate but equal." It is to ignore, as well, the Supreme Court decision in 1954—*Brown v. Board of Education of Topeka*—that effectively overturned *Plessy* and declared school segregation unconstitutional. It is to overlook a vast number of important legal cases that have shaped who was taught and what was taught—the "Scopes Monkey Trial" of 1925, for example, in which substitute teacher John Scopes was charged with violating a Tennessee law banning the teaching of evolution. It is to ignore the numerous changes to the public school curriculum, not simply involving controversial topics like evolution or sex ed: the addition of vocational classes like typing and "shop" and the removal of classes, too—the elimination of German language instruction from many schools, for example, following the country's entry into World War I. It is to overlook the launch of the Soviet satellite Sputnik in 1957, just one of many events that has prompted widespread handwringing about the state of public education. It is to skip over wartime and postwar pressures, including the influence of McCarthyism, to mold and enforce students' patriotism, to curb teachers' radicalism, and to monitor intellectual pursuits. It is to gloss over the establishment of fire drills, atomic bomb drills, and school shooter drills and the changing expectations of how schools need to keep their students "safe." It is to brush off the addition of healthcare and social services to schools and to discount the budget cuts that have since eliminated many of these. It is to ignore the rise of intelligence testing and standardized testing in the 1920s—long, long before No Child Left Behind, President George W. Bush's signature education law and the piece of legislation most closely associated with standardized testing today. It is to ignore other important legislation too, like the GI Bill, the National Vocational Education Act, and the National School Lunch Act. It is to discount the student protests of the 1960s and the changing legal status of students, particularly with regard to their First Amendment rights. It is to overlook the changing demographics of who attends school and who graduates and who goes on to college—not just the raw numbers or percentages, which have ticked up dramatically since 1892, but the composition of the public-school student body in terms of race, ethnicity, language spoken at home, socioeconomic status, and (dis)ability; it is to minimize how students themselves constitute and change the system. It is to deny that there have been many, many efforts to alter educational policies and rethink educational practices—at the K-12 and the university level—with varying degrees of success. The Carnegie Unit. The Life Adjustment movement. Progressivism. Open classrooms. Small schools. Phonics. The New Math. Back-to-Basics. AP classes. Ethnic studies. STEM. "Everybody should learn to code." And on and on.

Of course, Sal Khan is hardly the only person to push a narrative that schools have not changed in a hundred or so years. It's a popular story, one heavily favored by those who call for school reform. As the information technology sector has become more financially and politically powerful in the last decade or so, the voice of Silicon Valley has grown louder in the debates about the shape and direction of the education system. Many of its entrepreneurs have launched or invested in education businesses, often proudly ignorant of the history of education or the history of education technology. Steeped in what British media theorists Richard Barbrook and Andy Cameron have called "the Californian ideology"—"a mix of cybernetics, free market economics, and counter-culture libertarianism"—these technology-oriented reformers eschew the careful study of the past, preferring the kind of shorthand evident in the "History of Education" video.[5] "The only thing that matters is the future," one entrepreneur commented. "I don't even know why we study history. It's entertaining, I guess—the dinosaurs and the Neanderthals and the Industrial Revolution, and stuff like that. But what already happened doesn't really matter. You don't need to know that history to build on what they made. In technology, all that matters is tomorrow."[6] If you can peddle the story that everything was stagnant until you came along, your ideas, your inventions might seem that much more innovative and necessary—or so you hope.

Even if business executives and politicians care little or know little about history, they still tend to have a rough sketch of the past in their heads. Whether that sketch is accurate or not, they build products and outline policies and predict the future based upon it. They use history "willy-nilly," as historian David Tyack put it.[7] The task then is to help make sure more people get the story right.

One of the great flaws (and perhaps the great ironies) about that giant leap Sal Khan and Michael Noer make in their video, when they jump from the Committee of Ten to the invention of the internet and the work of Khan Academy, is that it ignores the long history of education technology. (Despite Khan's claim in his TED Talk that to "use video to reinvent education" is a novel idea, classrooms have been using film for over one hundred years.) As such, it erases the work of other education technologists who, throughout the twentieth century, built machines that did exactly what Khan claims Khan Academy will now for the first time in history do: enable students to move at their own pace through course materials, to receive immediate feedback on their progress, and to move forward to the next lesson only when they have demonstrated mastery of the topic. What today's technology-oriented education reformers claim is a new idea—"personalized learning"—that was unattainable if not unimaginable until recent advances in computing and data analysis has actually been the goal of technology-oriented education reformers for almost a century. Education psychologists like Sidney Pressey, the person often credited with inventing the first "teaching machine," talked about using mechanical devices in the 1920s in ways almost identical to those who push for personalized learning today, all so that, as Pressey put it, a teacher could focus on her "real function" in the classroom: "inspirational and thought-stimulating activities," including giving each student individualized attention.[8]

This book is the history of those first teaching machines.

But *Teaching Machines* isn't just a story about machines. It's a story about people, politics, systems, markets, and culture. It's a story of the twentieth-century education technologists and education psychologists and education publishers and education reformers who built and sold (or at least tried to build and sell) machines they claimed could automate self-instruction, could engineer a more personalized—or as they were more likely to call it, "individualized"—education system. It's a story of how education became a technocracy, and it's a story about how education technology became big business. It's a story of how the *science* of teaching and learning, as well as our *imagination* about teaching and learning, came to be caught up in mechanization, in efficiency, and, to quote the French philosopher Jacques Ellul, in "psychopedagogic technique."[9]

With *Teaching Machines*, I want to correct the historical record, if you will, to flesh out that period on Sal Khan's quickly sketched timeline between the Committee of Ten and the internet—in part to show that personalized learning is not, in fact, a "hot, new thing," invented by and facilitated by the latest batch of technology startups and the technology billionaires who fund them. Nor is resistance to a mechanized education just a matter of technologically backward teachers, clinging to a system that refuses to change. Indeed, some of those who resisted and undermined the development of teaching machines in the 1950s and 1960s were education reformers and corporative executives—the very figures who claim they'll fix things for us today. To correct the historical record then is to show there is a long history of automating and individualizing education and to demonstrate that much of today's education technology has its roots in these mid-twentieth-century teaching machines.

To tell the story of teaching machines, this book draws on archival research—on the letters and memoranda sent by educational psychologists, publishers, and machine makers. It traces the development of the technology through scholarly journals and government reports. It pays particular attention to the role of the media—newspapers, magazines, television, and film—in shaping people's perceptions of teaching machines as well as the psychological theories underpinning them. This "ed-tech imaginary" has blurred what we think about the past and the future. Indeed, it has, in many ways, been more influential than the history or science of teaching and learning. This imaginary has been instrumental in shaping people's beliefs in what technology can and should do in school.

I aim to challenge another element of the popular narrative about education, what I call "the teleology of ed tech."

There is a certain inevitability to the way in which education technology is pitched and packaged. One has no choice but to accept that schooling—and society at large—will become more technological, more "data-fied," more computerized, more automated. Resistance to this fate has kept education chained to its moribund methods, so we're told. To borrow a phrase from *Star Trek*, "resistance is futile." Even if, as the popular narrative would have it, the school system has remained unchanged for centuries, the digital classroom is imminent, and the computational future for teaching and learning is inescapable.

That those who work in and invest in education technology believe in education technology should come as no surprise. By definition and design, the field is inseparable from technology, inextricably tied to one technology in particular: the computer.

The computer, along with the internet, is often posited as the pinnacle of the history of education technology—"the end of history" in the very Francis Fukuyama sense, where computer-assisted instruction is the final, exultant form of education.[10] Certainly that's the story that Sal Khan told to *Forbes*. The computer and the internet are triumphant, and there's no going "back."

When I started work on this book, I sketched an outline that, unintentionally, replicated much of this type of historical predestination—a story where the machines got better, or at least faster, as the decades went by. The implication—one that I am certainly not comfortable in making: the instruction got better too. I first imagined that *Teaching Machines* would open with the multiple-choice teaching machines designed by psychologist Sidney Pressey in the 1920s and would describe his failure to successfully manufacture and market the device. It would then explore the work of B. F. Skinner—arguably the best-known proponent of teaching machines—and the others who designed and tried to sell teaching machines in the 1950s and 1960s. It would move on to the computerized version of these teaching machines—the "computer-assisted instruction" or CAI of the 1970s and 1980s. And it would end with the work of Seymour Papert, the MIT mathematician who in the late 1960s helped develop LOGO, a programming language for children, and whose views on teaching and learning were the antithesis of the proponents of both mechanical and computerized teaching machines.

"You can't stop there," one colleague protested. "You have to talk about Microsoft and Apple too! You have to talk about Steve Jobs and Bill Gates!" "You can't stop at the Eighties," another writer advised. "What about the Internet? What about 'mobile'?" And yes, the story could stretch on: what about massive open online courses, adaptive learning, predictive learning, and AI-enhanced tutor-bots—all more recent developments that could certainly be described as new versions of the "teaching machine."

Despite my insistence that today's education reformers, entrepreneurs, and technologists need to get a better grip on history, I'm not sure it's necessary to write another lengthy compendium of all the theories and tools utilized in educational settings from the Sophists on. Paul Saettler was one of the first to do this in his 1,500-page dissertation in 1953, a massive project that was shortened to some 400-odd pages when it was published as *A History of Instructional Technology* by McGraw-Hill in 1968.[11] Few books in the field reach that length these days (thank goodness), but the tendency to try to catalog every technology ever adopted by schools remains.

Saettler's 1968 book had, no surprise, very little to say about the computer. An updated edition, published in 1990 with the title *The Evolution of American Educational Technology*, included some 200 additional pages to address the new information technologies of the 1970s and 1980s.[12] And this is the impulse of many writers today, not just the authors of books but the creators of the endless lists and product directories. The "sweep of time," as Khan called it in his *Forbes* video, extends forward to include all the latest gadgets and gizmos, as though these necessarily are exciting or effective developments.

As a result, there's increasingly less space to discuss what happened prior to the computer. Indeed, anything that occurred before 1980, it seems, is treated simply as a precursor to the computer, only interesting or relevant insofar as it points toward the superiority of the new machines. One is meant to feel some narrative sense of relief, no doubt, that the computer has come along. Finally, education can make some progress.

Even by 1968, when the first edition of *A History of Instructional Technology* appeared, Saettler described the mounting criticism to B. F. Skinner's contributions to the field; and while the book was, as its title stated, a history, Saettler was prepared to talk about the prospects for new advancements—"systems engineering" and a "cybernetic analysis of the learner," for example—that would supplant what had been one of the most significant and heavily promoted developments of the previous decade: the teaching machine.[13]

This is certainly how the teaching machines of the mid-twentieth century are most often depicted: as a brief episode in education technology history, one that came to an abrupt end because the computer was looming on the horizon. Thus, teaching machines are depicted as failures—commercial failures if nothing else. This is the teleology of ed tech, where not only is the computer inevitable, but this notion of technological progress is the sole driver of events. It is as though the history of education technology, to borrow from *Wired Magazine* founder Kevin Kelly, is necessarily "what technology wants."[14] The success or failure of a technology, so this story goes, is because of technology itself, not because of high prices or poor quality or government regulations or changing social values or priorities.

In fact, there were decades between the teaching machine craze of the early 1960s and any significant or widespread adoption of computers in schools, and new technologies remain an area of resistance to as much as triumph for the computing industry.[15] Moreover, teaching machines and the pedagogy that accompanied the devices—"programmed instruction"—weren't just a phase or a fad. The ideas behind these developments—breaking lessons down into small, "bite-sized" content for instruction and assessment; the insistence that this would foster "individualization" in education by allowing students to move forward and master concepts at their own pace—were picked up by textbook publishers before being adopted by the early advocates for computer-based instruction. Even without long-term commercial success for the teaching machine makers, their ideas about programmed instruction have become "hard-coded" into all sorts of educational technologies and pedagogical practices.

The story of teaching machines isn't entirely unknown. B. F. Skinner, the name most closely associated with the devices, was arguably one of the best-known scientists of the twentieth century. As such, teaching machines are often decried these days as outdated behaviorist technologies, a disparagement that tends to overlook how much of Skinner's ideas—"conditioning," in his terms, or "nudging," in more recent Silicon Valley parlance—have made their way into the classroom via our contemporary computing devices.

Teaching machines and the psychological and pedagogical principles behind them are worth examining in their own right. Instead of writing a sweeping narrative that, inadvertently or not, positions the educational psychology of the 1950s and 1960s as merely a forerunner of the educational computing of the 1970s and 1980s or a prelude to the classroom apps of today, my storytelling here focuses exclusively on those earlier decades. To understand the teaching machines of the mid-twentieth century is to understand the teaching machines of today.

"A teaching machine is simply a mechanical device for presenting to a student a succession of instructional items requiring some discriminative response," philosophy professor John Blyth wrote in 1960, "and providing the student with an immediate check on the accuracy of his response."[16] *Teaching Machines* is obviously about the machines themselves—the various wooden and plastic devices built for the purposes Blyth describes. But the book is also about issues and events beyond the machines. The story of ed tech isn't simply a story of *tech*—a story that moves from product to product, from so-called innovation to innovation. Nor is the story of teaching machines merely the story of the scientists and businesses who built and sold them or the students and teachers who used them.

Teaching machines are bound up in the twentieth-century faith in science and technology and a fascination with gadgetry. They're also tied to other socioeconomic and sociopolitical forces, such as automation, standardization, and individualization—in education and more broadly in postwar American culture. Teaching machines are part of the growth of an ed-tech market, revealing companies' long-running interest in (and skepticism about) selling their wares to schools. They're also intertwined with the rise of educational psychology as a field and the field's encouragement of psycho-technologies for popular consumption, not just for the university laboratory. Teaching machines are connected to the changing expectations of what the school curriculum should look like and how it should be designed and delivered. The development of these devices occurred alongside major shifts in industry, institutions, and society at large, including the education system itself. With these shifts came new expectations of who schools should serve: Students? Civil society? Business? National defense?

This book tells the story of the *American* education system in the twentieth century. I recognize that that will irritate some readers who (rightly) contend that discussion about technology, education or otherwise, is already far too focused on the United States. Teaching machines—and along with them a belief in the power of automation, the need for individualization, and the necessity of business involvement in education—reflect elements of American culture and the American school system. That's not to say that teaching machines were solely an American phenomenon. The Soviets also built teaching machines in the 1950s and 1960s, and there were teaching machine manufacturers in other countries, including the UK and West Germany.[17]

Too often, the context is stripped from the stories written about education technology, and all that seems to matter is the technology itself. Its history is simply a list of technological developments with no recognition that other events occurred, that other forces—cultural, institutional, political, economic, and so on—were at play. If you were only to read the histories of education and education technology as told by technologists and technology boosters, you'd end up, no doubt, with a story much like the one Sal Khan offers in his video—a story in which there is no mention of racial segregation or desegregation or re-segregation, no mention of protests over wars or civil rights, no mention of legislation or court rulings. The satellite Sputnik is granted more agency in shaping twentieth-century education than students or teachers.

The history of education technology matters. This book tells just one little piece of it—the story of teaching machines. It situates these machines in a history of educational reforms and instructional practices. It considers teaching machines alongside the academic disciplines responsible for their development, as well as the major political events that encouraged and (discouraged) teaching machines' acceptance. It examines the "business of education," particularly the rise of the powerful testing and textbook industries and these industries' role in stimulating and squelching a teaching machine market. It describes an economic climate that would lead manufacturers to reject and embrace and reject teaching machines in turn, not only as a product to sell but as a method to train their own workers. *Teaching Machines* scrutinizes the role of the press in promoting teaching machines. It looks at how teaching machines were depicted in popular culture, as well as how their most well-known proponent, B. F. Skinner, was both hailed and pilloried for his science and technologies of social engineering. And it highlights the reactions to teaching machines by the growing youth and civil rights movements of the 1960s, asking a question that remains relevant today: how might "programmed instruction" and "personalized learning" enhance or impede freedom?

1

B. F. SKINNER BUILDS A TEACHING MACHINE

In the fall of 1953, Harvard psychology professor B. F. Skinner visited his daughter's fourth grade math class at Shady Hill, a private school in Cambridge, Massachusetts, where he observed the teacher and students with dismay. The students were all seated at their desks, working on arithmetic problems written on the blackboard as the teacher walked up and down the rows of desks, looking at the students' work, pointing out the mistakes that she noticed. Some students finished the work quickly, Skinner reported, and squirmed in their seats with impatience, waiting for the next set of instructions. Other students squirmed with frustration as they struggled to finish the assignment at all. Eventually the lesson was over; the work was collected so the teacher could take the papers home, grade them, and return them to the class the following day.

"I suddenly realized that something must be done," Skinner later wrote in his autobiography.[1] The classroom activities violated two key principles of his behaviorist theory of learning. Students were not told *immediately* whether they had an answer right or wrong. A graded paper returned a day later failed to offer the type of prompt and positive feedback that Skinner believed necessary to modify behavior—that is, to learn. Furthermore, the students were all forced to proceed at the same pace through the lesson, regardless of their level of ability or comprehension. This method of instruction provided the *wrong sort* of reinforcement, Skinner argued, penalizing the students who could move more quickly as well as those who needed to move more slowly through the materials.

A few days later, Skinner built a prototype of a mechanical device that he believed would solve these problems—and solve them not only for his daughter's classroom but ideally for the entire education system. His teaching machine, he argued, would enable a student to progress through exercises that were perfectly matched to her level of knowledge and skill, assessing her understanding of each new concept, and giving immediate feedback and encouragement along the way.

It was a "primitive" machine, Skinner admitted, fashioned out of a rectangular wooden box (see figure 1.1). "Problems in arithmetic were printed on cards," he explained. "The student placed the card in the machine and composed a two-digital answer along one side by moving two levers. If the answer was right, a light appeared in a hole in the card."[2] He quickly built a second model in which a student manipulated sliders bearing the numbers 0 through 9 in order to compose an answer. In another prototype, the student turned a knob after setting the answer. If the answer was wrong, the knob would not turn. If the answer was right, the knob would move freely, and a bell would ring.

Figure 1.1
Image of Skinner's teaching machine is openly licensed and available via Wikipedia at https://en.wikipedia.org/wiki/Teaching_machine.

A ringing bell is associated with some of the earliest and most famous experiments in behavior modification, namely those of the Russian physiologist Ivan Pavlov. Pavlov had published his research on dogs in 1897, describing how he'd conditioned the animals to respond to a bell by salivating—work for which he would later win the Nobel Prize in Medicine.

Skinner's own research was, to a certain extent, built on Pavlov's, moving from what was considered the "classical conditioning" of involuntary responses—stimulating salivation with food, for example—to an "operant conditioning" of voluntary ones. By using operating conditioning—that is, by administering rewards or punishments—all sorts of behaviors could be manipulated, Skinner argued, not simply reflex responses; and these behaviors could be bolstered through "schedules of reinforcement," the title of Skinner's 1957 book cowritten with colleague Charles Ferster.[3] Although Skinner insisted that he and Pavlov "were studying very different processes," the Russian scientist was incredibly influential on the early science of learning in general, focused as it mostly was on animal rather than human behavior.[4]

For Skinner, studying learning meant studying behavior, and vice versa. "For me," he wrote in his autobiography, "behaviorism was psychology."[5] Skinner contrasted this with "mentalism," a belief to which he would frequently accuse his fellow students and professors of ascribing. By "mentalism," Skinner meant both Freudian and Jungian analysis—that is, ideas about consciousness and unconsciousness, ideas that had garnered significant popular not just scientific appeal in the early twentieth century. One could not observe or verify what happens in "the mind," behaviorists like Skinner contended, and therefore "the mind" itself could not really be examined through scientific experimentation or inquiry. Indeed, in reviewing Carl Jung's book *Psychological Types* in 1923 for the *New Republic*, behaviorist John B. Watson—arguably the best known American behaviorist before Skinner—dismissed the work of the Swiss psychoanalyst as relying on "unjustified and unsupportable assumptions," on "magic" and not science.[6] Scientific study, behaviorists insisted, meant analyzing activities—*behaviors*—rather than speculating about inward motivations or sensations.

Skinner described his approach as *radical* behaviorism, which he argued "does not deny the possibility of self-observation or self-knowledge or its possible usefulness, but it questions the nature of what is felt or observed and hence known. . . . The position can be stated as follows: what is felt or introspectively observed is not some nonphysical world of consciousness, mind, or mental life but the observer's own body."[7] Rather than seeing "the mind" as entirely beyond scientific inquiry, Skinner argued that one could actually examine events "taking place in the private world within the skin"[8]—but one must do so through a behaviorist lens. This meant that language and learning, as historian of psychology Alexandra Rutherford points out in her work on Skinner's cultural impact, "all come under the purview of the experimental analysis of behavior, but they are radically reconceptualized as forms of behavior ultimately dependent on the external or social environment for their development."[9]

If behavior was controlled (and controllable) by the environment, then what better way to make adjustments to individuals—and, as Skinner imagined, *to all of society*—than by machine.

Skinner's commitment to behaviorism was not simply "academic," a term that is often used to suggest a theory divorced from practice. Skinner was a best-selling author, a public intellectual, a "visible scientist."[10] He was an inventor of psychological gadgetry and a promoter of what Rutherford has called "a technology of behavior"—a technology that, despite Skinner's rather controversial reputation, has "become a clearly identifiable component of life beyond the laboratory."[11] Indeed, Rutherford argues that "Skinner's most enduring cultural legacy is his technology of behavior, rather than his experimental science or his philosophy of radical behaviorism."[12]

And that is a legacy that is foundational for education technology. It's not where the story of teaching machines begins, but it's almost always how the story of teaching machines ends: deeply intertwined with Skinner and with his psycho-technologies. It is a foundation from which education technology has never entirely broken.

As one of the leading scholars in the field of behavioral science, Skinner was invited to speak at the University of Pittsburgh in March 1954 at a conference on the practical applications of behaviorism, and "excited about teaching machines," in his words, he decided to use the opportunity to discuss education. With his machine on a platform beside him, he presented to his peers a paper titled "The Science of Learning and the Art of Teaching," which was published that summer in the *Harvard Educational Review*.[13]

Skinner opened his remarks by chronicling some of the recent achievements in the field of learning. "Special techniques have been designed," he explained, "to arrange what are called contingencies of reinforcement—the relations which prevail between behavior on the one hand and the consequences of that behavior on the other—with the result that a much more effective control of behavior has been achieved."[14]

Skinner used his speech to review for the audience his research on pigeons—work that had begun during World War II. (His earliest experiments—experiments that had led to the invention of a training device dubbed the "Skinner Box"—had been performed on rats.) Skinner had built a contraption that, using his conditioning methods, would enable a pigeon to steer a flying object; and during a war increasingly reliant on air weaponry, Skinner imagined that that object could even be a missile. In 1943, General Mills had been awarded a $25,000 contract to develop a "homing device" under the code name Project Pigeon.[15] But the military never implemented the pigeon-guided missiles, and Skinner wrote in his autobiography that "Project Pigeon was discouraging."[16] Other scientists and engineers were not convinced the idea would work—they "simply did not trust pigeons," he admitted, insisting that "it was not a total loss."[17] "Our pigeons never had the chance to be heroes," Skinner later lamented, but they had "established themselves as excellent laboratory subjects."[18]

Project Pigeon helped Skinner to recognize that his research "was no longer merely an experimental analysis. It had given rise to a technology."[19] His work with the pigeons brought him widespread recognition in both the academic and popular press, even as many of the details of Project Pigeon remained classified. "Pigeons Play Piano and Do Other Smart Things," the *Worcester Gazette* had reported about those "smart birds at Harvard" a few years earlier, showcasing the birds from Skinner's lab that had been trained to play ping-pong.[20]

"From this exciting prospect of an advancing science of learning, it is a great shock," Skinner told the crowd at the University of Pittsburgh, "to turn to that branch of technology which is most directly concerned with the learning process—education."[21] Skinner then proceeded to repeat his complaints about the arrangement of the classroom, noting that unlike his pigeons, which had been taught to perform tasks by being rewarded with food pellets, most students experienced negative rather than positive reinforcements. Behavioral management had become less a matter of corporal punishment in schools, he admitted, but "anyone who visits the lower grades of the average school today will observe that a change has been made, not from aversive to positive control, but from one form of aversive stimulation to another. The child at his desk, filling in his workbook, is behaving primarily to escape from the threat of a series of minor aversive events—the teacher's displeasure, the criticism or ridicule of his classmates, an ignominious showing in a competition, low marks, a trip to the office 'to be talked to' by the principal, or a word to the parent who may still resort to the birch rod."[22]

Positive reinforcement was possible, but Skinner estimated that it would take "something on the order of 25,000 contingencies" to properly shape mathematical behavior—"a guess" he later admitted in his autobiography.[23] Of course, a classroom full of students all relied on the same teacher to provide this feedback, and a teacher could not possibly provide that many corrections or consequences.

"The simple fact is that, as a mere reinforcing mechanism, the teacher is out of date," Skinner told his Pittsburgh colleagues. "This would be true even if a single teacher devoted all her time to a single child, but her inadequacy is multiplied many-fold when she must serve as a reinforcing device to many children at once. If the teacher is to take advantage of recent advances in the study of learning, she must have the help of mechanical devices."[24]

"He's kidding," a professor of education in the audience muttered to a colleague.[25]

He wasn't.

His teaching machine, Skinner argued, boasted several important features: it gave immediate feedback, and while a whole class could set to work on their devices, each individual student could progress at their own rate. Furthermore, the machine could liberate the teacher from the tedious work of grading. "Marking a set of papers in arithmetic— 'Yes, nine and six *are* fifteen; no, nine and seven *are not* eighteen'—is beneath the dignity of any intelligent person," Skinner asserted.[26]

A brief account of Skinner's machine appeared in July 1954 in the *Science News Letter*, similarly concluding that the device would aid student learning, and with it, "the teacher is relieved of the time-wasting, temper-trying task of correcting stacks of arithmetic papers."[27]

Newspapers picked up the story that summer too. "Machine teaches kids arithmetic painlessly," the *New York Herald Tribune* reported. The *Worcester Telegram* announced the "Machine Age in Teaching." The *World Telegram* called it "the Atomic Age." "Miracle Gadget Makes Boys Like Arithmetic," the *Boston Herald* declared.[28] The stories all credited the Harvard psychologist with the invention of the device. The teaching machine seemed just the right sort of gadget to stir postwar America's imagination about the future of education.

A few years later, in the December 1957 issue of *Contemporary Psychology*, its editor, the renowned historian of psychology Edwin G. Boring, invoked Skinner and his role in inventing teaching machines in his monthly (gossip) column, CP SPEAKS. Following a paragraph praising the publication by the University of Pittsburgh's Lloyd Homme and David Klaus of a manual that would aid students in running their own behavioral conditioning experiments, Boring made a brief aside: "Homme, by the way, has been at Harvard for the past year, working on another idea that originated with Skinner, the technique for teaching by machine."[29]

Boring, who'd been quite cautious in his own scholarship about making these sorts of categorical claims of origin or invention, should probably have known better than to call Skinner "the first."[30] But the shaping of the origin myth of teaching machines was already well underway—until, that is, the action of Ohio State University (OSU) psychologist Horace English who wrote, rather indignantly, to Boring on January 30, 1958.

English said he felt obliged to correct Boring's history and his attribution of the invention to Skinner: "When the public press gave Skinner credit for originating 'teaching by machine,' it did not seem worth while to protest. But when the scholarly editor of CP falls into this trap; dear, oh dear." English stepped up to defend his OSU colleague, Sidney Pressey.

> Be it known then that Sidney Pressey published descriptions of machines which gave reinforcement by immediate knowledge of results and of one machine which gave reinforcement in the form of a piece of candy—when the child pressed the right lever, if you please. There have been at least six doctoral dissertations devoted to finding the effectiveness of such machines. And there is even a published bibliography of such devices, including many experimenters who preceded Pressey into the field. Most of this was before our friend Skinner even finished grade school.

"Originate, indeed!" English huffed, then quickly added, "This note is not for publication but it should lead to some sort of correction. Sidney hasn't said beans in my hearing;

maybe he does not know of Skinner's new proposal. Does Skinner have any idea of the wealth of research that has preceded him?"[31]

Boring promptly (and rather sheepishly) wrote back to English. "While it's a nuisance not being omniscient, I remain much more interested in education than in defense of the acts of this fallible organism that I call Me. Error often promotes truth, and this error could produce a good note on the history of teaching machines, using CP to belabor."[32] Boring said he would ask Pressey to write an article for *Contemporary Psychology* to clarify matters.

Two days later, Boring had lunch with his fellow Harvard colleague Skinner and grilled him about the history of teaching machines, a conversation that can't have been too pleasant since, decades earlier, Boring had criticized Skinner's account of the history of psychology and had challenged several passages in Skinner's dissertation: "I fear that you may be distorting history," Boring had written in the margins. (Skinner boasted in his autobiography that he'd refused to make any changes to his text in response.[33]) During their lunch, Skinner informed Boring that he was indeed familiar with Pressey's work; moreover, he was in the middle of writing an article that would tackle the long history of teaching machines.

"I suggested to Fred that he, instead of Sidney or Horace, pick on me for the education of the public," Boring wrote in a letter to the two Ohio State psychology professors later that day, "that he send CP a statement about the history of teaching machines and I will run it as a letter or in CP SPEAKS, depending on what it looks like when I see it."[34]

In June, a correction did finally appear in CP SPEAKS—not a statement from Skinner, but rather a short bit by Boring, opening in the typical, punchy manner of the editor's column: "Here comes Horace B. English, on his familiar charger, right to the door of CP's tent, defending the honor of Sidney L. Pressey who invented a teaching machine before Skinner had even got to Hamilton College."[35] The column, while mildly apologetic to Pressey about the error, still centered Skinner in the narrative. Boring cited Skinner's position that "he doubted whether the originator of the teaching machine could now be identified," pointing to a patent filed in 1886 by one Halcyon Skinner (no relation to B. F.) for a machine for teaching spelling. Boring tried to clarify too—an indication, perhaps, that Skinner had helped in crafting this response after all—that there were important differences between the machines of Pressey and those of Skinner: "The teaching of Pressey's machine depended on trial-error-and-success with the successes reinforced," the column read. "Skinner is undertaking to substitute operant responses, eliminating errors. Nothing but success."

Boring admitted in a letter to English and Pressey that he did not "fully understand the difference."[36]

Later that year, in October, Skinner published his article on teaching machines in *Science*—an article which, as he'd described it to Boring, did give credit to Pressey for designing devices in the 1920s. But much as in Boring's CP SPEAKS column, Skinner insisted that *his* theory of learning—his practices of behavioral management and arrangements of "contingencies of reinforcement"—differed significantly from Pressey's and as such, the two men's teaching machines differed as well. The science of learning had advanced much in the decades since Pressey's invention, Skinner maintained, although the practices of schooling had not. The "'industrial revolution in education' which Pressey envisioned stubbornly refused to come about," Skinner observed. "Pressey's machines succumbed in part to cultural inertia; the world of education was not ready for them."[37]

Arguably, Pressey shared some of that sentiment: his attempts to commercialize teaching machines in the 1920s and 1930s had indeed not gone very far. Despite his disappointments, however, he had remained attuned to the field. He had seen the short account of Skinner's teaching machines in *Science News Letter* in 1954 and had subsequently written to the Harvard professor that year, sending him reprints of the three articles he'd published in *School and Society* on the topic (in 1926, 1927, and 1932) and encouraging him to "stay with the general idea and see it through to some general fruition."[38]

"It has long been my belief," Pressey told Skinner, "that mechanical devices in schools were as feasible as in banks, and when I saw your pigeon demonstration at the Cleveland [American Psychological Association] meetings last year I wondered if something of this sort was not in the offering. I shall expect to see a busy child into [*sic*] a similar display case in New York this September." Skinner had lunch with Pressey and his wife that fall at the annual APA conference— "the most stimulating episode of the meetings," Pressey later wrote to Skinner, full of encouragement for him to continue to pursue the development of teaching machines.[39]

A few months after their New York meeting, Pressey wrote to Skinner again, this time with a much more somber and cautionary tone:

> May I urge, however, that in advance of any manufacture, you do continue experimentation. An experience of mine over twenty years ago may have some relevance. I then got an apparatus company to manufacture, and put on preliminary sale, one of my devices. It soon became evident that I had not got all the bugs out of the mechanism. Also, I had not got all of the psychological bugs out of the idea. In successive runs through the machine, students did indeed learn very rapidly, and so far everything looked encouraging. But then, and also in more recent experimentation, when I checked on that learning by repeating some of the material verbatim but in a little different context a week later, very little appeared left. If the test was rephrased and in a different context, almost nothing seemed retained. As indicated in the reprint I sent you recently, it was only when our devices became part of a total system of instruction that retention seemed enough to be possibly worthwhile. Of course, I expected losses over a period of time, and in transfer. But though I was not unsophisticated in experiments in learning, and of long experience in class-room experimentation, I was still surprised at the smallness of the outcomes.[40]

Be wary of overpromising when it comes to the science, Pressey warned. Be wary of overcommitting when it comes to the business. The advice from one professor-inventor to another: be wary of how you proceed.

Skinner, however, was quite convinced that his machine was superior, that the learning science was better, and that the education world was, in fact, finally ready for automation. Moreover, he had the deep connections to industry and to financing that his professorship at Harvard provided. Francis Keppel, the dean of the Harvard Graduate School of

Education, had already made an introduction to a member of IBM's board of directors. Skinner was in contact with "the most prestigious firm of patent lawyers in Boston."[41] Surely this time, the "industrial revolution in education" that Sidney Pressey had predicted decades before was about to come to pass.

Surely this time, things would be different.

2

SIDNEY PRESSEY AND THE AUTOMATIC TEACHER

"The future of civilization is well nigh in your hands," President Calvin Coolidge declared on the last day of 1924 as he welcomed a group of three thousand scientists to the White House. "You are the wonder workers of all ages. The marvels of discovery and progress have become commonplace. . . . Those of us who represent social organization and political institutions look upon you with a feeling that includes much of awe and something of fear as we ask ourselves to what revolution you will next require us to adapt our scheme of human relations."[1]

The scientists were gathered in Washington, DC, for the biennial joint meeting of the American Psychological Association (APA) and the American Association for the Advancement of Science. The president's remarks, covered the next day on the front page of the *Washington Post*, underscored a faith in progress that characterized not only how politicians and the press but also how the public were to welcome the twentieth century's scientific and technological advances—faith and perhaps, as Coolidge confessed, "something of fear." The invitation to the White House granted an important legitimacy to the assembled group, particularly to those from the relatively new field of psychology, now sanctioned as "wonder workers" by the highest office in the land.

Professor Sidney Pressey attended the APA meeting with several of his colleagues from Ohio State University, which boasted one of the most influential psychology departments in the United States—a department, in the words of its chair, "at grips with the central purpose of reducing education to a science."[2] At the meeting, Pressey, who had joined the OSU faculty in 1921 but was not yet a full professor, delivered a paper and demonstrated his prototype for a "simple machine for automatic testing of intelligence or information."[3]

Intelligence testing had become one of the predominant means by which psychology sought to "reduce education to a science," making its systems more scrutable and its outcomes more quantifiable. One of the earliest and largest applications of intelligence testing in the United States was not in schools, however; it was in the military. Harvard psychology professor Robert Yerkes had developed a standardized test called the Army Alpha, which, beginning in 1917, was administered to some two million recruits in order to identify their capacity for serving and their potential for leadership. Sidney Pressey, who'd been a graduate student of Yerkes at Harvard, had helped develop the testing items and apparatus, and Yerkes had recommended him for the Army's Psychological Corps, where many of the nation's top psychologists would serve during World War I. However, in August 1917, four months after the American entry into the conflict, Pressey was classified as "physically deficient" by the military due "marked myopia" and a "systolic murmur."[4] As such, he was disqualified from serving at all, let alone earning a prestigious military commission. Pressey remained a professor at Indiana University during the war, where he conducted research at hospitals and schools, administering "mental surveys" of children in the region.

By 1920, the standardized testing of students was already commonplace. This standardization was meant to ensure that the content, administration, and scoring of tests were objective—something that the largely male profession of psychology seemed to doubt the largely female ranks of schoolteachers could ever be without their scientific intervention. There was simply too much "arbitrary personal opinion" in schools, one administrator at the time decried, and "standards of excellence" were clearly necessary.[5]

Testing had quickly become a thriving industry, and vendors were "circling the world with psychological supplies," as one article in the 1927 edition of *Industrial Psychology Monthly* crowed.[6] University psychologists actively worked with (and worked *as*) businesses to develop and sell new products. Sidney Pressey was quite representative in this respect, and he had found a great deal of success peddling his standardized tests to schools. He and his wife Luella Cole had forty-seven different tests on the market.[7] Millions of copies of these had been sold, along with thousands of copies of their textbook, *Introduction to the Use of Standard Tests*, which they marketed to teachers and principals in order to encourage them to adopt this new type of assessment.[8] "The blanks cost only $1.25 per hundred," Pressey later wrote in an autobiographical essay—an amount that, with inflation, is about $18 in today's dollars. "A class could be tested in twenty-five minutes, blanks scored in one per minute," Pressey maintained.[9] What schools paid for in these testing materials, they were promised to make up for in saved labor costs.

In 1915, Pressey had sketched an initial design for a machine that would automate the grading of these tests, further hastening the pace of evaluation and giving the test taker and test administrator an immediate score. The war prevented much work on the idea, but he eventually built a prototype out of typewriter parts.

That machine, which Pressey exhibited at the 1924 APA meeting, administered a multiple-choice test: "What does perjury mean?" read one question. The possible answers: "(1) Lying (2) Swearing (3) Slander (4) Gossip."[10] The test was fed into the machine on a sheet of paper just as one would load a piece of paper into a typewriter. The test taker had four keys with which to respond, and after selecting her answer, the machine would advance automatically to the next question, calculating the number of correct responses along the way. Alternately, a lever in the back could change its operation slightly, and the machine would not move on to the next question until the test taker got it right, tabulating the number of tries on each question.

In February, about a month after Pressey returned home from the Washington, DC, conference, a local Ohio paper covered his invention: "By simply pressing a key, the person tested revealed his mentality or lack of it."[11] Based on the feedback he received from his fellow psychologists, this interest from the media, and his previous experience in the testing business, Pressey was confident he could bring his "Automatic Teacher" to market.

One year after first demonstrating the machine to his peers, Pressey began to look for a business partner—someone who could help him build and distribute the device. He wrote to publishers. He wrote to the manufacturers of mimeograph and adding machines. He wrote to the makers of typewriters and cash registers. He reached out to the suppliers of various kinds of scientific equipment, including those producing polygraph machines, another new apparatus that had emerged from the psychology laboratory around the same time. In letter after letter, Pressey tried to convince potential manufacturers that the fabrication of his teaching machine would be simple and cheap—between $50 and $60, he estimated in one letter—and that there would be a strong demand from universities and schools alike.[12] (This is the equivalent of $700 in today's dollars.)

"The device is simpler than an alarm clock, and is made of sheet brass," Pressey explained in a letter to the A. B. Dick Company in December 1925, adding helpfully, "the parts could be stamped out."[13] "Many parts of the device are typewriter parts," he wrote to the Remington Typewriter Company in April 1926, "and the remainder of the machine composed largely of flat pieces which would be stamped out. Manufacture of the device by such a company as yours would thus be a very simple and inexpensive proposition."[14]

Pressey pitched the teaching machine as a reflection of the demand, if not the necessity, for technical and scientific progress in education. Automation was inevitable—or so he hoped. As he wrote to Remington, "I am not alone in believing that extensive use of mechanical aids in education is a probable development of the near future. If such a development does come, it will not be a negligible matter commercially."[15] Teaching machines, Pressey argued, were poised to become an incredibly lucrative product, and a manufacturer would be wise to help him establish the market.

Some companies responded graciously to the professor, thanking him for drawing their attention to the invention and remarking it was "very interesting" and "very clever." Others were quite dismissive in their replies.[16] "We are not interested," the president of Underwood Typewriter sneered, "and fail to see where this would interest a typewriter manufacturer. The device has no types nor is it in any way related to typewriting, being outside of our line we would not care to consider anything in connection with this device."[17]

The Dalton Adding Machine Company responded to Pressey in April 1926, indicating that, after some consideration, it *might* be willing to develop the machine. But its list of conditions was long: Pressey would need to entirely redesign it to the company's specifications, and he would have to personally foot the bill for a new punch press and screw machine necessary for its manufacture—all at the cost, Dalton estimated, "in the neighborhood of $10,000" (approximately $140,000 today).[18]

The president of the Marietta Apparatus Company, A. G. Watson, answered Pressey's inquiry in February 1927, and he was one of the few to take the idea seriously as a business *and* pedagogical proposition, responding not only positively, but at some length. Watson wrote to Pressey that he felt the teaching machine needed to meet three criteria in order to be commercialized successfully: it needed to be inexpensive; it needed to adapt to a variety of situations; and it needed to require little time or effort for educators or students to use.[19]

"I feel very sure," Watson wrote, "that the biggest selling point in your device is the labor saving, drudgery eliminating service it will render in connection with the ever present duty of testing." Watson said that he'd spoken to a junior high school teacher who seemed "very enthusiastic" about the idea of the machine. The teacher had told Watson that he began each class with a "snappy" true-false quiz and then had his sister grade them. "The machine would eliminate the kid sister," Watson admitted, but that was perhaps a good thing. Indeed, Watson wrote, Pressey's teaching machine helped to underscore that "the need is felt for frequent testing but the burden of scoring is prohibitive."

And yet, even though he penned several pages of enthusiastic feedback on the mechanics of the device and on its potential application in classrooms, like many of the others who responded to Pressey, Watson admitted that in the end "my problem is financial, not scientific."

Arguably, Pressey's problem was financial too, and in January 1928, he began to seek meetings with a number of regional banks, hoping to secure financing that might give him more leverage in his negotiations with manufacturers. He asked for twenty minutes with the Cleveland Trust Company, where Dr. Leonard Ayres was vice president. (Ayres had authored an incendiary book in 1909 on the dismal state of US education, *Laggards in Our Schools*, and had served as a statistician for the US Army in World War I.[20]) "I think that I ought not to let you expend some of your time in talking to me about it," Ayres responded disdainfully. "I cannot just now devote any time or effort to my old educational efforts."[21]

Although Pressey had substantial experience with the education business when it came to the publishing of standardized tests and textbooks, the manufacturing of a piece of scientific equipment was something quite different; and Pressey failed at first to recognize this distinction in several important ways. He wrote to some manufacturers offering to negotiate with them on the royalties he'd earn on sales, for example—a standard financial bargain *authors* struck with publishers—but he had made no attempts to secure a patent, the legal framework that protects *inventors'* intellectual property. Manufacturers were very clear they were unwilling to develop "any invention which is not thoroughly protected."[22] Pressey filed the paperwork for his first patent, the "machine for intelligent tests," in June 1928. (The patent was not awarded until March 1930.)[23]

Alongside his dizzying campaign of letter writing to potential business partners and investors, Pressey continued his academic research on testing and teaching machines, publishing the paper he'd delivered at the 1924 APA meeting—"A Simple Apparatus Which Gives Tests and Scores—and Teaches"—in the March 1926 volume of the academic journal *School and Society*. This article, which focused primarily on how his machine functioned, concluded by articulating Pressey's major argument for its adoption: labor-saving mechanisms of this kind would benefit schools. Pressey contended that, unlike almost every other type of work, educators had failed to pursue automation, in part Pressey claimed, because educational institutions were so conservative. These institutions were unwilling to address their inefficiencies because "a teacher's time is cheap."[24]

Schools were hardly as technologically backward as Pressey implied in his *School and Society* article, however. By the

1920s, public education had already undergone several decades of reforms that encouraged schools to operate much more like businesses. Although the idea of compulsory schooling was still relatively new, a flood of immigration at the turn of the century had placed enormous pressures on the US education system. Schools were blasted as wasteful, full of students who were "laggards" and teachers who were failing to train their pupils academically or morally. Few politicians were eager to allocate more money for education even with the swelling enrollments; instead, they demanded more efficiency. That is, they wanted to institute some version of "scientific management," a concept that had become popular for factory operation, in order to make schools more standardized and more cost-effective.

The work of Frederick Winslow Taylor, a mechanical engineer, had captivated the imagination of American business leaders in the early 1900s with his call to apply scientific principles—namely analysis, measurement, and improvement—to the workplace in order to boost efficiency. Taylor's book *The Principles of Scientific Management* was published in 1911 and within a year was being applied to almost all aspects of American life, well beyond the production line.[25] As Raymond Callahan documents in his 1962 book *Education and the Cult of Efficiency*, Taylor's ideas were quickly brought to education: "Scientific Management and High School Efficiency" was the topic of a High School Teachers' Association meeting in New York City in 1912; "Improving School Systems by Scientific Management" was the theme of the 1913 convention of the Department of Superintendence (now the National Association of School Superintendents).[26]

For schools, as for factories, scientific management meant the strict control of inputs and outputs; monitoring, analyzing, and planning all aspects of the job; setting standards so that workers were kept at the ideal pace on the ideal task. The key, for Taylor, was to conduct a time and motion study. For schools in particular, this meant adoption of many new procedures for measuring and accounting for what went on in the classroom. According to Ellwood P. Cubberly, dean of the School of Education at Stanford University and author of the influential 1916 treatise *Public School Administration*, the need for efficiency demanded the testing and classification of students: "Our schools are, in a sense, factories in which the raw products (children) are to be shaped and fashioned into products to meet the various demands of life. The specifications for manufacturing come from the demands of twentieth-century civilization and it is the business of the school to build its pupils according to the specifications laid down. This demands good tools, specialized machinery, continuous measurement of production to see if it is according to the specifications, the elimination of waste in manufacture, and a large variety in the output."[27]

This push for "good tools" and "specialized machinery" could be seen in the kinds of products that schools readily procured in the 1910s and 1920s: standardized tests obviously, but also textbooks, workbooks, and flash cards. Some schools also adopted the very latest broadcasting technologies of the time—that is, film and radio—all in the name of instructional efficiency.

Unlike the "mass education" of the radio or the film projector, Pressey claimed his Automatic Teacher would foster a more *individualized* classroom. He recognized that "some sentimentalists" would resist "education by machine."[28] But he insisted that machines would actually free the teacher "from the mechanical tasks of her profession—the burden of paper work and routine drill—so that she may be a real teacher, not largely a clerical worker."[29] He offered an example of how his own department operated, hoping to make the case for how much time and money could be saved by automating these practices:

> The writer is in immediate charge of a first course in educational psychology which is handled by a number of instructors. In this course it has been the custom to give each week a short quiz—usually of the objective-answer type. These quizzes take about five hours per section to score. There is an average of five sections per quarter for the four quarters of the year. The total time cost of scoring these tests is thus for the year about 1000 hours, or 125 eight-hour days—about five months (21 weeks) of one person's full time. Thirty-five of the above-described simple machines would have saved practically all this labor.[30]

His teaching machine, he insisted, would "leave the teacher more free for her most important work, for developing in her pupils' fine enthusiasms, clear thinking, and high ideals."[31]

What Pressey seemed to overlook, of course, was how much his own work and his own profession, by promoting practices like standardized testing, had contributed to these mundane working conditions in the first place.

In May 1927, Pressey published a follow-up article in *School and Society*: "A Machine for Automatic Teaching of Drill Material." Although quite similar to the first one that had appeared in the journal, Pressey focused this time around less on the machine's capacity to reshape teaching labor and more on how the Automatic Teacher exemplified the very best science and science-informed pedagogy.[32] Pressey drew on the work of Columbia University professor Edward Thorndike, one of the most influential psychologists of the time, to argue the teaching machine reinforced the "laws" of learning that Thorndike had famously developed.

According to Thorndike's "law of recency," for example, the most recent lesson was the one best recalled. Pressey argued that because his machine functioned so that the last answer was always the correct answer—that is, the one that advanced the machine on to the next question—the student would better retain the material. And if a student struggled to get the correct answer, the repetition would help further retention as well—Thorndike's "law of exercise." The machine would reinforce the right kind of learning behavior (and whether they admitted it or not, almost all psychologists at the time were behaviorists) and do so in a pleasing and positive manner—Thorndike's "law of effect."[33] Pressey even added an optional feature to the design of a machine that would dispense a candy whenever the student got an answer right.

After the publication of the *School and Society* article, University of Michigan psychology professor (and developer of a very popular arithmetic test) S. A. Courtis wrote to Pressey, thanking him for planting "a very large 'seed' thought" and predicting that teaching machines would "have a very remarkable development over the next fifty years."[34] Pressey wrote back to Courtis, confessing that investors did not seem to share that excitement and that several manufacturers had stopped answering his letters altogether. "Things are thus largely at a standstill," he admitted. "I am lying awake nights from impatience to develop the fascinating possibilities which it seems to me may open up"—if only someone would appreciate the financial and educational potential of his teaching machine.[35]

Finally, in early 1929—after three years of mailing inquiries and receiving almost twenty rejection letters in return—it appeared as though someone would: the W. M. Welch Manufacturing Company.

Based in Chicago, W. M. Welch Manufacturing had been founded in 1906 by a former school superintendent, William Welch, whose first commercial product in the late 1800s was a record-keeping and award system for schools—the diploma.[36] By the turn of the century, the company had a thriving business building and selling furniture and laboratory supplies to schools and hospitals as well, and as such, a solid base of education customers. (Welch told *Time* in 1939 his company was selling over half a million diplomas a year.[37]) Although Welch had a good reputation, Pressey was hesitant that his teaching machine would be treated like "a piece of freak apparatus," and he volunteered to help with the marketing "through professional contacts and demonstration-use of the device in summer school classes made up largely of school administrators."[38] "I am sufficiently confident in this matter," Pressey wrote to William Welch, "that I am willing to include in any contract we might write a clause providing that I am to market the first 200 of the machine, the only requirement being that (for these introductory copies at least) the price be not over $5.00. And if the 200 can be ready in good season for summer school I am willing to add that these sales should be made by the end of the year." ($5 is about $70 in today's dollars.)

After years of failing to make any progress with manufacturing the machine, Pressey pressed forward with his new business relationship with Welch Manufacturing with great enthusiasm, signing an initial agreement with the company in January 1929, submitting an order for thirty machines for his summer course at Ohio State that same month, and even proposing that these not count toward the number of sales required before he'd see any royalties. A few months later, increasingly desperate to get the production underway, Pressey offered to pay $1,200 toward the tooling costs for designing the manufacturing components—about $17,500 today.

The company did not share Pressey's impatience. Sales manager (and eventually president) Medard W. Welch wrote to Pressey in May, indicating that the company "distinctly disliked" to accept Pressey's personal funds. "We would prefer to make the mailings, solicit the orders in your name and after we are assured that production can be undertaken without personal loss to you or any loss to our company, we will proceed with the work," Welch wrote.[39] He sought to reassure Pressey that "a slower and more conservative approach" would be preferable, and Pressey need not worry as his patent application would protect his financial interests no matter the pace of production. Welch insisted the company remained very enthusiastic about Pressey's idea and about their business relationship, but he invoked his company's lengthy experience selling various products to schools—Pressey would need to trust him. Welch said that he had observed how challenging it was to market anything that required teachers to alter their classroom practices. "In numerous instances where pieces have had distinct merit we have found extreme resistance among school people to changing [the] method of doing the work," Welch explained. "In other words, the personal view enters into teaching to such a great extent that it is difficult to make such projects move rapidly."[40] "Reducing education to a science," as Pressey's department head at Ohio State University had called it, required first convincing teachers to change their practices—and that was not an easy sell.

By the end of May, Welch Manufacturing had created a circular advertising "a New Automatic Testing Machine for Teaching and Testing," with a price tag "not over $15" (approximately $220 in today's dollars). It wasn't quite the same as having a machine in hand, but the advertisement seemed to indicate that production was imminent. Pressey was elated. "Immediately upon receiving the printed announcements," Pressey wrote to Medard Welch in June, "I sent out nine hundred covering the membership of three national organizations, the American Psychological Association, The Educational Research Association, and the National Association of College Teachers of Education. . . . There are people interested in the device all the way from Yale to Stanford, and almost all the big universities are interested."[41]

"You certainly are a genius in perfecting such a contrivance," Stanford University's Ellwood P. Cubberly effused to Pressey in response to receiving the circular.[42] "I feel quite sure that we are on the eve of very interesting and valuable developments in the line of test scoring and tabulating machines," wrote Columbia University professor Ben D. Wood, head of the school's Bureau of Collegiate Educational Research and someone keenly interested in the automation of test scoring.[43] Emboldened by the praise from his peers

Figure 2.1

Promotional flyer for Sidney Pressey's Automatic Teacher used with permission from Ohio State University Archive.

and by the sales that started to trickle in, Pressey urged William Welch, the company's president, once again to speed up the manufacturing so that the devices could be delivered in October. Welch agreed to not wait for further orders, and he said he would put production in line. "To do this to best advantage," he wrote to Pressey in June, "we ought to have your working model, and suggest that you return it to us, by express collect, as soon as possible."[44]

But even as Welch prepared to begin the manufacturing process, Pressey continued to tweak its design, working over the summer months with an engineer at OSU to refine various pieces of the apparatus. He peppered the company with letters and with detailed suggestions for changes: add a snap-in sheet-metal bottom, adjust a spring mechanism, move a lever, change the size of a pawl, alter the ratchet wheel.[45] "We want to be sure everything is all right," he fretted in a letter dated September 17, 1929. "It would be fatal if everything were not."[46] He wrote to the company again the next day, with still more suggestions, requesting to see a revised production model. "I hope you can" send it, he ended his letter, repeating "as I want to be sure everything is all right."[47]

Welch Manufacturing's executives responded calmly to Pressey's flurry of letters, thanking him for his input and noting that none appeared to require any major alterations to the production tools. But Richard Welch, the company's production manager, confessed, "We are having a terrible time to make the tooling inexpensive. Time after time we have found cheap fixtures impractical and consequently sunk from 50% over our original estimate for tools for the part in question to more than double the amount. There is only one possible way that I see relief can be obtained: namely boost the sale high."[48] Welch warned Pressey that the company was poised to be out between $500 and $1,000 just on the production of the first one hundred machines. (This is roughly $7,000–$14,000 in today's dollars). The company would have to sell 250 just to break even. Welch urged Pressey to do everything he possibly could to boost the sales.

Pressey responded to Welch's pleas for him to focus on the marketing with a letter that instead focused on the engineering, certain that a smooth-functioning machine would do more to drive sales. "Those who get one now will (if the first model works entirely satisfactorily) want more," he insisted.[49]

In the beginning of October 1929, the company shipped Pressey a model made almost entirely from the production tooling. "From this point on," Richard Welch stated firmly in the cover letter that accompanied the sample machine, "only vital mistakes in construction can be altered."[50] But Pressey was not satisfied at all and sent the machine back with two pages of further feedback. A pawl didn't engage properly. A spring was not strong enough. There was too much play in the keys. Pressey traveled from Columbus to Chicago, visiting the production plant personally to supervise the changes.

The flaws with the production model persisted—at least according to Pressey. When he received another sample device in November, he discovered that there were problems aligning the key sheet, and as such the machine would not consistently register the user's responses.[51] Pressey proposed several modifications to address this, but the company was reluctant to make any changes at all as these "would make the tool cost very much higher."[52]

October 1929 brought "The Great Crash," as share prices on the New York Stock Exchange collapsed, sending the economy into a tailspin. Pressey's personal health was also headed in a dangerous, downward spiral. He collapsed and was ordered by his physician to remain "more or less inactive" for the month of December.[53] Yet even from his sick bed, he continued to write letters to Welch Manufacturing, punctuated excessively with exclamation points and with suggestions—some new and some repetitive—about the machine. Pressey was worried about being left out of discussions, panicked about the delivery date—a date that Welch would not commit to—and concerned the company would not be able to fill existing orders by the end of the calendar year. Any fears about a looming depression went unmentioned in the correspondence between the professor and the manufacturer. Perhaps that's because Pressey was not obsessed with the market; he was obsessed with his machine.

Pressey spent the next eight months in and out of a sanitarium. In spite of this, he continued his work, which frenetic letter writing aside, was still, first and foremost, the work of a college professor: teaching, research, administration. During the spring term of 1930, Pressey taught three courses at OSU and supervised five class sections. Between 1928 and 1930, he advised twelve master's degree and ten doctoral students. "The plain truth is that both the Presseys have been working at top speed for two or three years," the head of his department George Arps wrote concernedly to the university president in January 1930. "I have cautioned them from time to time and made available as much clerical and laboratory assistance as was within my power to give them. Two years ago I cautioned them of the day of reckoning which would be sure to come."[54]

In early February 1930, Richard Welch sent a letter in care of Pressey's wife, Luella Cole, suggesting she only share it with the professor if he was healthy enough to receive the news it contained. "It is not as optimistic as I would like to write."[55] The letter informed Pressey that it would be financially impossible for Welch Manufacturing to make any more changes to the teaching machine. The company was not publicly traded, but the stock market crash had affected its business nonetheless, and Richard Welch was increasingly skeptical that the teaching machine was worth the investment. "I have tried to avoid harping on the subject," he wrote, "but we are out a great deal of money on the initial quantities of these testers. They are running much farther beyond the estimates than I had anticipated on account of the additional die costs. As this is true I don't want to add one penny of unnecessary experimental work until the maximum return is secured."[56] Pressey was free to tinker with the machine on his own time, Welch said, but the tooling had been cast, and the company was ready to begin production with the model it had on hand.

In April, Welch Manufacturing shipped 100 machines, and Richard Welch wrote to Pressey stating he was confident that an additional 150 would be ready to ship by June.[57] Thirty of those first machines went to OSU to fulfill the order the university had made for them back in January of the previous year. Pressey quickly dispatched one of his graduate students, Lyle Addie, to investigate how effective the machines worked in comparison to paper-and-pencil assessment, and Addie concluded that "tests given by the machine are substantially as reliable as tests given in the usual way, and measure substantially the same thing."[58]

Pressey confessed to one colleague that his illness had "almost demoralized my work."[59] But with the machine poised to finally hit the market, his spirits were buoyed, and he was again certain the teaching machine would become a sensation. And it was—briefly and locally at least.

"The labor saving machine is advancing upon the schoolroom," the *Columbus Dispatch* reported in April.

> What the calculator is to the office, the electric sweeper to the home, a simple apparatus which tests, scores, and also helps in teaching, may be to the schoolroom through the efforts of an Ohio State university professor. The teacher too may now be freed from certain drudgeries through the use of labor-saving devices, just as the homemaker, the office and

factory worker. . . . His small device, which looks much like a diminutive adding machine, is the pioneer in the application of laborsaving mechanisms to education, the latest outpost to be affected by the machine. . . . In certain respects, no human teacher can compete with its nicety and precision.[60]

A student-run publication, the *Ohio State University Monthly,* also covered Pressey's invention, although with a much wryer tone, quipping that if someone could just invent another machine that would punch the right answers into Pressey's teaching machine, the future of education would be "perfect in the eyes of the student." The article criticized a university culture that had become so focused on test taking and on cheating. Test taking wasn't just drudgery for teachers; it was drudgery for students too. And it was quite clear, to OSU students at least, that taking tests via machine would not address that at all.[61]

Although his moods swung from elation to despair, Pressey continued to work on the design of the machine. Over the summer months, he also sought to elicit more scholarly feedback and transform some of the media attention into sales. Medard Welch informed Pressey in June that Welch Manufacturing had mailed some two thousand letters to college professors and normal school instructors, advertising the machine. The company had sold eighty and had another twenty out on commission.[62] Pressey presented the Automatic Teacher at the National Education Association's annual meeting in July, and he was invited to exhibit the machine at the 1933–1934 World's Fair in Chicago.

But the problems with the Automatic Teacher persisted, specifically in two areas: price and functionality. The $15 price tag was beyond the reach of almost all schools, and as Pressey himself pointed out, that amount was almost half the cost of the per-pupil expenditure in 1930.[63] Much of his argument for the device had rested on its labor-saving potential. But as the Great Depression began to take hold, local tax revenues fell, and school budgets were slashed. In districts across the country, teachers were laid off; in some, teachers worked without pay. Although enrollment expanded as teens opted to stay in school rather than join the ranks of the unemployed, there was little to no appeal of a machine that could replace or even supplement the work of a teacher.

The problem with the Automatic Teacher wasn't just the economy. As Pressey had feared all along, many of those who had purchased a teaching machine found problems with its operation. Pressey wrote with some frustration to Welch in June, sharing with the company the negative feedback he'd received from S. A. Courtis. "Dr. Courtis is one of the two or three most important men in the country in experimental education and in tests."[64] The operational issue that Courtis and others experienced involved the way the machine recorded test answers, just as Pressey had suspected based on the final sample he'd examined.

Each test required two sheets of paper: a key sheet and the student's answer sheet. The former, a copy of the test printed on heavy paper stock with holes punched in it to indicate the correct answers, had to line up correctly with the latter for the mechanism to register if the answer was right or wrong. If the sheets were not aligned properly, the machine could not accurately record a student's score—if it was able to register anything any all.

Pressey urged Welch Manufacturing to make whatever improvements Courtis wanted, offering to "make any adjustment as regards to royalty or otherwise which might seem desirable." He was desperate. "If we could get Courtis active in behalf of the machine, that would help a great big lot." What he didn't write but was surely thinking: if Courtis was vocal in his criticism of the machine, everything would be ruined.

Richard Welch had come to believe that any flaws were with Pressey's initial design rather than with his company's manufacturing. But he pushed back nevertheless on Pressey's claim that customers were truly dissatisfied. "There has been very little trouble reported," he wrote to Pressey in October 1930, dismissing the complaint as an issue with how the devices were being *used* rather than how they had been *built*.[65] That said, Welch did recognize that the key sheets could become a problem.

Pressey proposed that perhaps uncut key sheets be included with the machine, allowing the test administers to punch the key to match their test sheets themselves. Welch thought this was a bad idea—"the average person will not take the time and care for hand punching sheets with a degree of accuracy that will make them infallible."[66] Welch suggested instead that key sheets, punched to give four different patterns for true-false tests, be included with each machine. This meant, of course, that the answer patterns for the standardized tests to be used with the machines would be standardized themselves, as educators would have to make their tests conform to the pre-set answer sheets that came packaged with the Automatic Teacher.

Even with the publicity Pressey had tried to drum up over the summer months, by the end of October, only 127 machines had been sold—less than half what the company had anticipated—and Welch still had sixty-nine on hand. As such he was unwilling "under any circumstance" to invest more money in the device until sales warranted it. Production, he informed Pressey, would cease immediately.[67]

Pressey was furious. "May I speak frankly?" he fired back in an angry letter, which, undated, was possibly never sent. In it, he unloaded his frustrations and disgust at the company:

> Your shop made dies for this machine before it understood how the device worked; and when I first arrived in Chicago to check up on a model which your shop sent me which would not work, found that no one in the shop had taken the trouble to determine the nature of the mechanism. . . . At a cost of strength and health which I do not care to contemplate I managed to patch up the thing and further the sale so that you disposed of 160 machines. If your shop had done a reasonably good job three times as many might have been marketed simply to psychological laboratories, for laboratory use, without reference to possible use in schools.

Pressey charged that the "hostility" of Welch Manufacturing had damaged both of their reputations. "The whole business has been unfortunate to you; it has been exceedingly unfortunate to me."[68]

Even decades later, Pressey remained angry. "The Welch Company made a mess of the construction of the testing machine," he wrote to a colleague in 1941, "and I literally made myself sick at the time about the whole business."[69]

Nonetheless, Pressey continued to write letters to Welch, proposing ideas for a new and updated machine, one that would be simpler, one that would be cheaper, one that would be more easily manipulated by young children. He seemed unable to admit failure. He seemed unwilling to abandon the idea.

In August 1931, Medard Welch finally wrote back to Pressey, having let many letters go unanswered. He was as restrained as he could be. "Don't you think from the business stand-point, it would be preferable with the firms in the business, to find some way to disposing of the materials on hand before putting out a new model. . . . I had hoped very much that these machines might be sold prior to this time, especially after your lecture at Cleveland, but schools have apparently tightened up on their appropriations and we have got to look forward to disposing of the machines after the schools open up."[70] Welch was polite but adamant: the manufacturing of the machine was done.

Pressey responded to the end of the production of the Automatic Teacher perhaps the only way he knew how: he started writing letters to Welch's competitors. He reached out to other manufacturers—slot-machine companies and adding machine companies, as well as General Motors and the International Business Machine Corporation (better known by its initials, IBM)—trying to convince someone, anyone to go into business with him.[71]

In 1932, in his third (and final) article on teaching machines in *School and Society*, Pressey blasted education as "the one major activity in this country which is still in a crude handicraft stage."[72] Efficiency and labor-saving devices would have to be introduced, not to mechanize education but as a means of "freeing the teacher from the drudgeries of her work so that she may do more real teaching, giving to the pupil more adequate guidance in his learning." Pressey repeated his prediction that an "industrial revolution" in education was on the horizon.

But this revolution would not be Pressey's to fight, he declared. When it came to his own work on teaching machines, Pressey admitted defeat. "The writer is regretfully dropping further work on these problems," he wrote in a footnote to his final *School and Society* article. "But he hopes that enough has been done to stimulate other workers."

Pressey had failed to commercialize the Automatic Teacher. His attempts to do so had nearly destroyed his health, and it had likely contributed to the end of his marriage to Luella Cole. (The two divorced in 1933.[73]) Yet for the new field of education psychology and its related business of education technology, there would be no turning back from this work.

3

"MECHANICAL EDUCATION WANTED"

The Great Depression surely doomed Sidney Pressey's plans for the mass production of the Automatic Teacher. But it's possible that, even without the stock market crash, he'd have been unable to persuade typewriter manufacturers to build his teaching machine. They were already quite convinced that the typewriter itself could be the ideal, modern, educational device.

Another educational psychologist, Columbia University professor Ben D. Wood, had helped assure them of such. In 1932, he, along with University of Chicago colleague Frank Freeman, published a widely cited study that found typewriters in the classroom improved students' reading habits and increased the amount they wrote.[1] The typewriter experiment, which involved some fifteen thousand students in grades kindergarten through six, was "hailed as an original and important achievement in educational research," according to Wood's biographer.[2] At the very least, it received excellent press—a hint perhaps at how much good publicity counts as "achievement."[3]

An article in *School and Society*, for example, claimed that "children clamored for their turn at the machines."[4] This enthusiasm was shared by teachers as well, the article contended, and they reported that working with the machines encouraged students to be more cooperative and to develop more personal responsibility. By using typewriters, students could not only enhance their mastery of school subjects; they could also improve their attitudes toward school work in general. The article's author, the psychologist (and noted eugenicist) Albert Edward Wiggam, was optimistic that typewriters could end "the suffering we put children through in learning to write with pencil and pen—actual mental and physical suffering."[5]

This was precisely the hope and excitement that the typewriter manufacturers wanted to encourage. In the summer of 1928, the presidents of Remington Typewriter, Royal Typewriter, Smith & Corona, and Underwood Typewriter—among the largest makers of mechanical typewriters in the United States at the time—had held a meeting where they agreed to form an "educational bureau." "The object" of this industry-funded research organization "would be to obtain actual facts and experimental evidence as to the advantages of the use of the portable typewriter as an educational instrument."[6] The four companies agreed to contribute equally to the bureau, which, according to its initial arrangement, would not sell typewriters or be operated for financial gain. To that end, the companies contributed $32,000 to launch the Educational Bureau of Portable Typewriters in February 1929 and $76,000 to continue its work through 1930. (The total of $108,000 in 1930 is the equivalent of roughly $1.5 million today.) Ben Wood was hired to direct the organization and to lead the research into the effectiveness of the devices in the classroom.

Although they were underwriting the research, the typewriter manufacturers promised to abstain from directly marketing their products to the schools participating in the experiments or from using Wood or Freeman's names in their advertising materials. Nevertheless there were numerous episodes when salespeople did approach the schools—"bootleg typewriter activity," as the head of the Columbia University bookstore described some of the attempts to peddle products to the students and parents at Horace Mann School, a private school with ties to Columbia's Teachers College where some of the typewriter research was being conducted.[7] The companies were admonished for this behavior, but there were no real consequences. Wood said he felt it was his job, not the companies', to make sure the project maintained its integrity. It is "my responsibility," he wrote, "to the companies to protect their enterprise from any possible failure in the integrity of the experiment and in the way in which our report on the experiment will be received."[8] No doubt, what the typewriter manufacturers were interested in was less "the integrity of the experiment" and more the benefits— the *commercial* benefits, to be clear—they might accrue from what was a significant financial investment in the bureau.

After the initial flurry of good publicity from the release of the Wood-Freeman report, the typewriter companies became increasingly frustrated with the bureau's direction and the slow pace of adoption of typewriters by schools, most of which found making *any* purchases of *any* instructional devices utterly unfeasible during the Great Depression. As part of the organization's subsequent work, J. L.

Sweeney, who'd become managing director of the bureau in 1931, tried to shepherd the publication of a new typewriting curriculum through Macmillan, the publisher of the Wood-Freeman report; however, Sweeney eventually resigned following repeated confrontations with the typewriter executives, who balked at his insistence that they change the keyboard design and sell schools specialized machines. (Sweeney had made this recommendation based on the findings of Wood and Freeman, who had observed that small children were confused when they struck a typewriter key with a capital letter on it, only to have the lowercase letter appear on their paper.[9]) The majority of the sniping in the correspondence to the Educational Bureau of Portable Typewriters involved one manufacturer accusing another of improper behavior, as everyone involved was trying, if nothing else, to keep their heads above water during desperately difficult economic times.

Meanwhile, in Chicago, W. M. Welch Manufacturing had already ceased production of Sidney Pressey's Automatic Teacher.

Ben Wood along with Frank Freeman had undertaken a similar study for a different industry a few years earlier. In 1928, George Eastman, the cofounder of Eastman Kodak, had paid the two to investigate (and hopefully demonstrate) the educational advantages of film. Wood and Freeman designed a study of approximately eleven thousand students in grades four through nine that used twenty films integrated into two twelve-week units in geography and general science. Their findings were remarkably similar to those in their typewriter experiment: the students who were taught with films tended to score higher on end-of-unit assessments than those who were taught with printed materials. "The use of teaching films not only entails no loss in reaching the ordinary goals of education," Wood wrote, "but actually promotes their attainment to a significant degree, and that films can be made an integral and administratively feasible part of the regular school program, working with and enhancing the effectiveness of the customary pedagogical devices and procedures."[10]

Many "customary pedagogical devices and procedures" were already in place by the turn of the twentieth century. As historian Larry Cuban argues in his book *Teachers and Machines*, "By 1900, public schools had established organizational and classroom practices that would be familiar to present-day observers. Schools usually were divided into grades and separate classrooms, one to a teacher. Rows of desks bolted to the floor faced a chalkboard and teacher's desk (portable desks were installed in the 1900s but did not become common until the 1930s). . . . Report cards, homework, textbooks, teacher lectures, and student recitation were standard features of urban classrooms at the turn of the century."[11] As Cuban makes clear, a succession of technologies—textbooks, chalkboards, radios, televisions, typewriters—had also been introduced into classrooms around the same time to enrich teaching and learning and, just as importantly according to some proponents, to increase the efficiency of schooling.

As early as 1910, films were used regularly in many classrooms, even though they'd only been introduced to the general public just a decade or so before.[12] "I believe that the motion picture is destined to revolutionize our educational system and that in a few years it will supplant largely, if not entirely, the use of textbooks," Thomas Edison famously pronounced in 1922. "I should say that on average we get about two percent efficiency out of schoolbooks as they are written today. The education of the future, as I see it, will be conducted through the medium of the motion picture . . . where it should be possible to obtain one hundred percent efficiency."[13] Edison, much like George Eastman, was quite literally invested in films becoming embraced as a modern, education-technology intervention, poised to make students learn better and faster, with the same (or even less) teacher labor.

Industry was happy to pay for research that demonstrated as much. Critics of education in the early twentieth century charged, "instruction was regimented, mechanical, and mindless. Teachers, according to one researcher, told students 'when they should sit, when they should stand, where they should hang their coats, when they should turn their heads.' Students entered and exited the classrooms, rose and sat, wrote and spoke—as one," Cuban recounts.[14]

Ben Wood, like many educators at the time, opposed what he saw as these mechanistic elements in the school system, calling instead for the individualization of education. In an article published in the *Teachers College Record*, Wood blasted this standardization, invoking a speech that Harvard University president Charles Eliot had given in 1892 to promote the work of the Committee of Ten:

> Education is properly the development and training of the individual body, mind, and will; but when it is systematized, and provided for many thousands of pupils simultaneously, it almost inevitably takes to military or mechanical methods;

and these methods tend to produce a lock-step and a uniform speed, and result in a drill at word of command rather than in the free development of personal power in action. The interests of the individual are frequently lost sight of, or, rather, are served only as the individual can be treated as an average atom in a heterogeneous mass. This natural tendency in systems of education I believe to be a great evil, particularly in a democratic society, where other influences, governmental, industrial, and social, tend toward averaging the human stock.[15]

This was the conundrum for the school system: educate the masses, but resist standardizing them; expose everyone to the same curriculum, in part to "Americanize" them, but at the same time foster that core American value of individualism.

Wood argued that schools needed to become much more responsive to students' individual needs. But to do that, he contended, the system needed to move toward "mechanical education," something that was, to be clear, distinct from a *mechanistic* education. When he addressed the Harvard Teachers Association in 1931, he urged his audience "to make a serious effort to take full advantage of mechanical developments in making the school less costly to tax payers, more efficient and progressive from the viewpoint of society, and more pleasant and effective for the children."[16] By "mechanical education," Wood meant the use of scientific equipment—the typewriter, the phonograph, films, and the new giant calculators and computational devices, for example. To promote this, Wood served on the Committee on Scientific Aids to Learning, a subcommittee of the National Research Council (NRC), alongside other notable figures like Vannevar Bush, the head of the NRC; Frank Jewett, the president of the National Academy of Sciences and chairman of the board of Bell Laboratories; and James Bryant Conant, the president of Harvard University.

Perhaps this seems counterintuitive: to individualize education, one must automate it. To resist mechanistic education, schools must mechanize. But for education reformers in the early twentieth century (as for those in the early twenty-first), it was a conundrum they managed to justify. Indeed, this contradiction gets at the heart of calls for "personalization" and is central to a vision—then and now—of a modern, high-tech, progressive learning experience.[17]

In 1935, Wood wrote to John Dewey, the philosopher whose name remains the most closely associated with progressive education, acknowledging how much the latter's work had influenced his own and citing a passage from Dewey's *The Child and the Curriculum*: "the child is the starting point, the center, and the end. It is he and not the subject matter that determines the quality and quantity of learning."[18]

Although Dewey's writings were immensely popular, Wood lamented that the school system had largely failed to adjust to reflect the philosopher's ideas. "As I survey the schools more than a quarter century after you wrote these sentences," Wood continued, "I find that the child is still neither the starting point, nor the center, nor the end of our educational organization. So far as my observation goes, this is true even in the so-called progressive schools to some extent at least, since all of them start with a predetermined curriculum, and most, if not all of them, seek to apply one uniform standard of achievement."[19] For Wood, the individualization of education meant the modification of curriculum and of pedagogy, rather than the expectation that the student adjust to the curriculum or to the traditional teaching and administrative practices of school.

As Wood articulated in an address to the Institute for Personnel Workers in June 1925, the individualization of education would require that teachers know their students thoroughly. "Not only the colleges, but the schools, from kindergarten to university, must realize that their first duty is not to *teach* but to *learn* students. To get accurate and significant information about students, and to record it in a way that will be available and meaningful and directive at each step in the educational ladder, is a duty fully coordinate with, and certainly prerequisite to the proper discharge of the duty to teach students."[20] To "learn students" was not a matter of cultivating interpersonal rapport with each one as much as it was a matter of developing a scientific profile and a statistical analysis of them. To *know* students, for Wood, meant to *test* students—via content examinations, psychological analyses, personality assessments, and intelligence and aptitude tests.

But the current (or even enhanced) testing practices alone would not be sufficient. Pedagogical practices had to change and respond to the data in turn. Wood wrote in 1934 that "the chief defect in the testing movement has been the neglect of building an adequate philosophy and system of using test results for effective and constructive educational guidance in the larger sense of the term."[21] And *that* system demanded a *machinery*.

Although Wood's career was deeply intertwined with the development of some of this machinery, he is rarely mentioned in the histories of *teaching* machines, an omission that obscures how a "mechanical education" has long meant the adoption of business machinery for administrative ends and how "personalization" and "individualization" have long meant data collection and analysis. Wood's name is much more closely associated with the testing movement, as the cofounder, along with the American Council on Education (ACE), of the Cooperative Test Service that later became the Educational Testing Service (ETS). "I have all along not considered myself a test maker or a test peddler, but a tactical practitioner who, like a medical practitioner, uses precision instruments which fit his purpose," Wood told his biographer Matthew Downey in 1964.[22]

Like many education psychologists, Wood served in World War I as part of the US Army's psychological corps, continuing his graduate studies when he was discharged in 1919 at Columbia University, under the direction of Edward Thorndike. His dissertation examined experiments the college was undertaking in using psychological testing for admissions purposes. In 1923, Columbia created a professorship in Collegiate Educational Research, and Wood was appointed to the position. His office soon became officially known as the Bureau of Collegiate Educational Research—"one of the first testing and guidance bureaus set up by a liberal arts college."[23] While much of Wood's initial work explored examinations and admissions at Columbia, the new bureau was commissioned to develop tests for other organizations as well—most notably a set of tests in physics, French, Spanish, and German to be used in the New York State Regents Examination. "For the first time in the history of American education," Wood wrote in 1925, "we shall have a comprehensive objective survey of public instruction in four subject matters on a state-wide basis."[24]

The Regents Examination was scored by hand—an enormous task. So when Wood and his bureau undertook an even larger project in 1928—"the Pennsylvania Study," sponsored by the Carnegie Foundation, in which some 27,000 high school seniors and 4,500 college seniors were given a battery of standardized tests—he wrote to ten or so companies that manufactured calculators and business machines, expressing his interest in any sort of device that might automatically score students' exams.

Most of the companies did not bother to respond. But one CEO contacted Wood directly: Thomas J. Watson, Sr., the head of IBM.

As Wood's biographer Matthew Downey tells the story,

> [His] first meeting with Thomas J. Watson is one of Ben Wood's fondest memories. It marked the beginning of a close, lifelong friendship. Watson was such a busy man that he told Wood when they met at the Century Club that he had only an hour to state his business. A secretary was posted to remind Watson when the time was up. Wood's ideas about the potential uses of machines in education, especially the possibilities of electronic data processing machines (a figment of Wood's imagination at that early date), so appealed to Watson that he kept Wood talking the entire afternoon. Every hour on the hour the secretary was waved away.[25]

Watson made Wood a consultant to IBM and gave him an annual retainer of $5,000—about $75,000 in today's dollars. The deal was incredibly lucrative to Wood personally, but also professionally. Watson loaned Columbia several truckloads of computing equipment to aid in Wood's research, and soon this equipment became a valuable asset to the university—used by the Bureau of Collegiate Educational Research as well as other departments. Wood and his team employed the machines for the statistical analysis of test scores, and Wood worked closely with IBM engineers to develop different mechanical scoring techniques. One of the first successes of this collaboration came in 1932 when an IBM tabulator was modified to score the Strong Vocational Interest Blank, a test that was used for student counseling but that was quite expensive to grade and analyze—"nearly $5 per student, even when using the hand-sorted punch card method."[26] (That is the equivalent of about $100 per student today.)

Use of the IBM tabulator enabled Wood's office to dramatically increase the number of tests it processed, expanding from 160 in 1932 to 3,105 in the second half of 1933 when the machine was in full use. Even so, Wood found that the IBM device was not ideal for scoring exams. It was unusable for many popular objective tests. It was too slow and still too expensive.

Wood imagined some sort of electronic device that could function much more quickly and cheaply, and he worked with IBM engineers on several possible models. One version used a scanning device to "read" the answers; another used "the analogue principle, in which the score was recorded on an ammeter in units of electricity when the electric circuits in the machine were closed by a graphite pencil mark on the answer sheet."[27] The latter seemed to be the most promising direction, although engineers found that the amount of electricity conducted—and thus the score registered by the ammeter—varied according to the darkness of the pencil mark. (One can standardize a test, but it's much harder to standardize how hard a student will press down with their pencil to answer the questions.)

This problem stumped the IBM engineers, but it had actually already been solved by a Michigan high school teacher named Reynold B. Johnson, who'd developed his own test scoring machine a few years earlier. Johnson had inserted high resistor units of 2,000,000 ohms into the tiny circuits closed by the pencil marks, something that boosted the total resistance so that any variation in the darkness of the pencil marks—500 to 5000 ohms—were no longer significant. His machine, he claimed, could be built for about $100—about $2,000 in today's dollars, but still at the time far cheaper than the cost of running the IBM tabulator as a testing machine.

After a story about Johnson's machine, the "Markograph," appeared in the *Chicago Tribune* in 1933, a regional IBM salesperson brought the invention to the attention of the company.[28] In early October, G. W. Baehne, who worked in IBM's research department, wrote to the Superintendent of Schools in Ironwood, Michigan where Johnson had worked, in the hopes of learning more about what Johnson had built. "It so happens that for some time we have been working with the Research Department of Columbia University in developing a grading machine along similar lines," Baehne said, "and we expect that our machine ultimately will be able to take care of an unlimited number of questions and will make it unnecessary for the pupil to punch out the answers. There may be a possibility that by combining some of the good features of your machine with what we have in mind, progress can be made more quickly."[29] Baehne asked for more

Figure 3.1
Promotional photograph of the Markograph used with permission from the Educational Testing Services Archives.

details and was curious as to whether the Markograph was commercially available (see figure 3.1).

The letter took some time before it reached Johnson, who wrote back, saying that "we are very proud of the help and service we shall be able to offer educational institutions, civil service bureaus and other organizations and the suggestion of combining features of our machine with yours is dependent upon a large number of factors."[30] Baehne responded quickly and with approval over "your willingness to cooperate with us," but he indicated that he had to speak to other IBM executives first before anything could proceed formally.[31]

Whatever willingness or eagerness Johnson might have had at the outset soon ran headlong into the lumbering IBM bureaucracy, which Johnson found uncommunicative—or at best, terribly slow to respond. When almost a month had passed since he'd heard from Baehne, Johnson wrote to IBM again. "We would appreciate word from you. If you have decided to drop consideration of the Markograph we would like to have you write us to that effect."[32] Baehne responded ten days later, indicating that "I turned your letter over to another department" and promising that Johnson would hear more shortly.[33]

But more time passed, and Johnson still had not heard back from IBM. At the end of November, he wrote to the company again, announcing that his company, The Electrical Test-Corrector Co., had

> definitely decided to place the Markograph in the hands of some national organization that is well equipped to handle a machine of this kind. Because of your delay we have assumed that your company is possibly not interested in cooperating with us. We have, therefor, begun correspondence with a large corporation manufacturing electrical goods and we hope to begin building business negotiations with them soon. We do not hesitate to say that we would prefer to place our test corrector in the hands of the International Business Machines Corporation since you are already interested in test correctors.[34]

The company did not view this as any sort of ultimatum, and IBM continued to stall Johnson and to send him mixed signals about its interest. One letter, in January 1934, said that "this device is too far removed from our present line of endeavor to warrant our undertaking at this time."[35] But one month later, a different person from IBM wrote to Johnson in such a way that suggested the company was considering the project after all, asking for more details about the patents he'd applied for and for estimates on the cost of fully automating the model and producing it "in lots of 100, 200, 500 and 1000."[36]

Johnson continued to try to go it alone, but with little success. In early March, he wrote to IBM again: "Since I have found it difficult to raise enough capital to warrant my attempting to market my machine and since I have likewise been unable to find a large organization to manufacture and undertake the sale of the machine at this time, I would like to have you consider the possibility of employing me in your research department to work with your experts in combining the good features of both of the machines."[37] Unlike those professor-inventors (like Pressey and Skinner) who could conduct their research and explore their business ideas from laboratories at prestigious universities, Johnson had no powerful institutional affiliation, and he recognized that his best bet might be to forego his own individual entrepreneurial efforts and join IBM. The Great Depression and the economic downturn had stretched on into the late 1930s after all.

At the end of March, G. W. Baehne, who'd assumed a new position in IBM's Educational Department, wrote to Johnson apologetically. "I may be in Milwaukee sometime during the end of next month," he wrote, "and shall make it my business to visit you and to study your machine at first hand."[38]

Baehne did not, however, make it to visit Johnson in person, so in April, Johnson offered to send his Markograph to IBM for study. Unfortunately, "upon opening the case containing the Markograph," Baehne informed Johnson in May, "I noticed to my regret that it had not survived the trip as well as I had hoped. The indicating needle on the ammeter was broken off, and as the device is of no use without it I was compelled to send it to the manufacturer for repair."[39] Despite the problems IBM faced in dismantling the device in order to repair it, Baehne told Johnson that the engineers liked the looks of it. Moreover, "there seems to be a reasonably large market for it."[40] But then, once again, the communication stopped, and Johnson had to pester the company again for more information on the status of its review of his machine. Baehne finally admitted that IBM still had not been able to get the machine to work, but he arranged for the machine to be delivered to Ben Wood's office at Columbia so that he could test it.[41] Wood was immediately impressed.

Although there were still some problems getting the Markograph to operate perfectly, Baehne told Johnson that he wanted a "definite proposition" that indicated what Johnson would require in order to sell patent rights to the machine to IBM. "If this proposition involves your personal services you should state your salary expectation and length of contract period."[42] Johnson responded that he would like to come to New York (at IBM's expense) so that he could personally demonstrate the machine to IBM executives. He indicated that "for full patent rights on the Markograph as shown in the present model and full patent rights on all improvements made . . . I have arrived at $15,000 as a fair price covering my ideas, time and money involved"—the equivalent of about $285,000 today.[43] Johnson also reiterated that he was interested in and qualified for research work with IBM. "Should I be contracted by you I would need $4,000 a year for not less than two years."[44]

As IBM continued its slow back-and-forth negotiations with Johnson, Wood finally intervened, writing directly to CEO Watson to praise Johnson's invention—and apologizing for interrupting Watson's vacation with his letter. "The basic principle of his machine," Wood wrote excitedly, "is so important that I am willing now to recommend the outright purchase of his main patents."[45]

On Wood's good word, IBM hired Johnson and, based on his Markograph, developed the first commercial test-scoring machine, the IBM model 805, which was unveiled in 1938.

Reynold Johnson was, no doubt, an excellent hire. He was a prolific inventor, and throughout his long employment by IBM—thirty-seven years—filed ninety some-odd patents, many relating to the storage and retrieval of data. Johnson helped develop, for example, IBM's first disk drive.

IBM's test scoring machine, however, was not a huge commercial success. Johnson estimated that only 1,000 or so devices were ever manufactured.[46] Nevertheless the machine had an enormous effect on the testing industry. "The revolutionary character of the influence of the improved scoring machines," Wood wrote in 1963, "not only on the testing movement but on our whole educational philosophy, has never been adequately explained or understood."[47] And certainly the prospects of mechanized scoring, as Wood's biographer argues, "permitted an unprecedented expansion of large-scale testing activities and huge reductions in the cost of testing."[48]

Johnson's machine was, of course, a testing machine and not a teaching machine. It's worth pointing out that that same charge was levied at Sidney Pressey's Automatic Teacher, a machine that is still credited as the first *teaching* machine. It's a charge that perhaps unintentionally highlights how closely the practices of testing and teaching had become—even in the 1930s and particularly with the introduction of this new educational machinery. Nonetheless, the importance of the IBM test scoring machine on the prospects for a broader teaching machine market and on the direction that all education technologies would take in the subsequent decades should not be overlooked. Just as Wood's work with the typewriter industry had likely decreased those companies' interest in manufacturing the machine that Pressey had designed, Wood's work with IBM seemed to convince the company that "mechanical education"— *individualized* education—was to be achieved through large-scale data analysis and *testing* machines. Even though the Markograph itself was not a great success, it helped convince IBM and others that there could be a substantial market for automated test scoring.

But just as importantly, as a product of IBM—the maker of business machines, the Markograph was built and sold as an *administrative* device—something quite distinct from Pressey's plans for a machine for classroom and laboratory use. Control of this new educational machinery would not be in the hands of K–12 teachers or students; it would be in the hands of principals and superintendents—in the hands of those who controlled the school budgets.

Ben Wood had insisted that there needed to be a stronger connection between testing and teaching, and as such, he believed that testing machines should shape pedagogical practices. And this meant in turn, no doubt, that testing machines would shape the design of teaching machines. Indeed, the Markograph did help "hard code," if you will, the multiple-choice question as one of the principles technologies of assessment in American classrooms, a practice that B. F. Skinner would come to reject in the design of his own, much better-known teaching machines.

But that meant too that Skinner would have to convince the corporate giant IBM to change the design of its machines—a next-to-impossible task and a good reminder that, despite the popular narrative, it is not necessarily *teachers* who offer the greatest resistance to changing pedagogical practices and technologies. Indeed, it is impossible to tell the story of teaching machines without telling the story of the *business* of education technology, without telling the story of corporations shaping what education technology looks like, how it functions, whose needs it serves. It is impossible to tell the story of teaching machines without telling the story of how corporations dragged their feet, slowed the development of products, stalled the market, resisted the latest science, and, in many ways, balked at educational change.

4

THE COMMERCIALIZATION OF B. F. SKINNER'S FIRST MACHINES

IBM was the first major corporation that showed an interest in manufacturing B. F. Skinner's teaching machine—but only after that famous Harvard alumni network was leveraged to get the company's attention. Francis Keppel, the dean of Harvard's Graduate School of Education, had seen a demo of Skinner's device and had been quite impressed. He'd told a Harvard officer about it, who suggested that Keppel reach out to Sherman Fairchild, a Harvard alum and member of IBM's board of directors. Keppel asked a classmate of Fairchild, Robert Gross, the CEO of Lockheed Martin, to make the pitch and to introduce Skinner's ideas and his machine to IBM. "What Professor Skinner proposes is nothing less than the mechanization of the American schoolroom in the teaching of arithmetic, spelling, and other drill subjects," Gross wrote to Fairchild, commiserating that "God knows the current situation in our schools is a distressing spectacle."[1]

In September 1954, Skinner received a letter from IBM expressing interest in his device and asking to forge a "confidential relationship" between the two parties. IBM requested Skinner sign a number of forms—nondisclosure agreements, in today's parlance—as "a necessary preliminary step."[2] Skinner balked. He wrote back to IBM stating that he'd be quite happy to explain his machines and theories to the company, but that he didn't feel like he should have to sign anything that would imply he had any "desire to submit ideas or inventions to IBM."[3] IBM replied that if Skinner planned to patent his teaching machine, discussions with the company could wait. Or, if Skinner preferred (and agreed in writing), IBM would assent to a meeting that would not be covered by any confidentiality protections—meaning both Skinner and IBM could pursue the business of teaching machines with other people or organizations.

Skinner agreed to the latter proposition and had his first meeting with IBM in November 1954. There, the two parties decided that IBM would explore the possibility of manufacturing a machine to teach spelling as well as arithmetic and that Skinner would try to find a philanthropic foundation to support the development of "programs"—that is, the instructional content to be used on the machines.

A few months later, in January of the new year, Skinner wrote again to his contact at IBM with some "afterthoughts":

> It seems to me that development of the machine and the preparation of the material to be used in it are almost inseparable and I am afraid this fact has some bearing upon the possibilities of finding support for both phases. I am not at all sanguine about getting a foundation to finance the preparation of the material if IBM is building the machines, not as a public service, but as a way of exploring a potential market for an eventual commercial model. I therefore very much hope that if IBM is interested in going ahead with this, that it

will feel that the machine itself is almost meaningless without the material and that it will therefore be willing to considering financing its preparation.[4]

A teaching machine, in other words, might become a profitable product, but its value—financially and pedagogically—was inseparable from the lessons designed for it. Skinner wanted IBM to develop the machine *and* the programs.

He assumed that IBM would move forward with this in mind, sending them tips on how the instructional materials should work: on how many "frames" should be in a single lesson—"25–50. About 200 lessons per term. To store material for whole term: about 5000 to 10000 frames"; on how a student's progress should be displayed; on how feedback for right and wrong answers should be instrumentalized, and so on.[5]

Skinner soon realized that IBM was less interested in designing a new machine, let alone in developing instructional materials, than in making one of its current products, namely the IBM Card Verifier, function as a teaching machine—at least while the company explored the feasibility (read: profitability) of programmed instruction. He understood that IBM might feel the need to conduct more testing before investing in the development of a brand-new machine. However, he told the company, if

> IBM shares my own strong conviction that recent advances in the field of learning promise great improvements in educational practices, provided certain necessary instruments can be made available, and that these improvements are not only desirable with respect to the general welfare but suggest a profitable field for commercial development, then I think that postponing the design of a more appropriate instrument for a year or more would be poor strategy. It is evident that

the field cannot be captured by patent rights. The first effective device on the market will have a tremendous commercial advantage.[6]

It was an argument Sidney Pressey had also made in an attempt to accelerate Welch Manufacturing's development cycle: first-mover advantage in a new teaching machine market would be crucial. Skinner pointed out that teaching machines were increasingly being recognized as viable technologies, particularly in the military, and that IBM's decision to rely on its Card Verifier for testing constituted a grave delay. It meant that "although we would get some idea of a few practical problems, we should have made very little progress toward a marketable machine or the most effective material to be used with it."[7]

By the end of November, Skinner wrote to IBM again, complaining that "more than a year has passed since IBM first asked for information about my work with teaching machines, and in retrospect I do not feel that very rapid progress has been made."[8] IBM needed to decide whether or not it was really interested in going into the teaching machine business, Skinner pressed. But just as Reynold Johnson had experienced a long and frustrating back-and-forth with the company, always holding out hope of working with IBM, Skinner relented a little and did not push too hard. He was very keen on a deal with the business machine giant. Its name recognition and its market dominance could be crucial for success, he thought.

The two parties agreed to a preliminary research and development schedule in December: by the fall of 1956, IBM would build ten models for machines that would teach arithmetic, and Skinner would prepare for a preliminary trial of the devices. IBM would install and maintain machines in schools by the spring of 1957 so that the devices could be tested. Preparation of materials and commercial production of machines would start that summer. The sale of the teaching machines would commence in the fall of 1958.

None of this was formalized, however, in any sort of written contract. A draft of one finally did appear in May 1956, which Skinner forwarded to his attorney, Paul Perkins, who revised it substantially—among other things increasing Skinner's remuneration.

When Skinner began these attempts to build his teaching machines in the mid-1950s, he was certain of two things. The first, as he wrote in his autobiography: "If teaching machines were to be used in schools, a company would have to manufacture them."[9] The development of the devices would require enterprise and capital, not just the expertise of researchers or educational institutions. And the second thing Skinner was sure of, based on his prior experiences pursuing the commercialization of one of his inventions, he would need a good lawyer.

A decade earlier, Skinner had attempted to bring another one of his psycho-technologies to market. After the end of World War II (and the end of Project Pigeon), Skinner and his wife chose to have a second child. "When Yvonne said that she did not mind bearing another child but rather dreaded the first year or two," Skinner wrote in his autobiography, "I suggested that we simplify the care of a baby"[10]—a process that became the topic for an article he published in *Ladies Home Journal* in October 1945, titled "Baby in a Box—Introducing the Mechanical Baby Tender."

"When we decided to have another child," he wrote in the magazine article, "my wife and I felt that it was time to apply a little labor-saving invention and design to the problems of the nursery. We began by going over the disheartening schedule of the young mother, step by step. We asked only one question: Is this practice important for the physical and psychological health of the baby? When it was not, we marked it for elimination. Then the 'gadgeteering' began."[11]

The crib Skinner "gadgeteered" for his daughter was made of metal, larger than a typical crib, and higher off the ground—labor-saving, in part, through less bending over, Skinner argued. It had three solid walls, a roof, and a safety-glass pane at the front which could be lowered to move the baby in and out. Canvas was stretched across the bottom to create a floor, and the bedding was stored on a spool outside the crib, to be rolled in to replace soiled linen. It was soundproof and "dirt proof," Skinner said, but its key feature was that the crib—an "air crib," Skinner called it—was temperature-controlled, so save the diaper, the baby could be kept unclothed and unbundled. Skinner argued that clothing created unnecessary laundry and inhibited the baby's movement and thus the baby's exploration of her world.

As a labor-saving machine, Skinner boasted that the air crib meant it would take only "about one and one-half hours each day to feed, change, and otherwise care for the baby." Skinner insisted that his daughter, who stayed in the crib for much of the first two years of her life, was not "socially starved and robbed of affection and mother love." Moreover, by dwelling in the crib, she'd avoided all colds and flu. "The compartment does not ostracize the baby," Skinner contended in the magazine. "The large window is no more of a social barrier than the bars of a crib. The baby follows what is going on in the room, smiles at passers-by, plays 'peek-a-boo' games, and obviously delights in company. And she is handled, talked to, and played with whenever she is changed or fed, and each afternoon during a play period, which is becoming longer as she grows older."[12]

The article's headline—"Baby in a Box"—was far from ideal, as it linked the crib to the animal training in Skinner's laboratory and with the "operant conditioning chamber" he used there, which was better known, even to the public, as the "Skinner box." The "Baby in a Box" could easily be read as less a new kind of crib built by an inventive father and more an experiment designed by a psychology professor.

The reaction to the article was overwhelming, and the incredible volume of correspondence Skinner received in response—immediately after the issue of *Ladies Home Journal* appeared on newsstands and for decades to come—was passionate no matter whether letter writers thought the invention was a technological wonder or an abomination.

This was, after all, postwar America, when increased consumption meant all sorts of gadgets were entering the home. Skinner recognized this, opening his article with a nod to the "brave new world which science is preparing for the housewife of the future."[13] As historian Karal Ann Marling has pointed out, "In the 1950s the United States bought fully three-fourths of all the appliances produced in the world," and these purchases were promoted as "a litmus test for the American lifestyle."[14] But there was a tension, psychologist Alexandra Rutherford contends, bound up in all this new instrumentation: "While being a good housewife in 1950s America meant participating in the consumer culture by purchasing appliances, tranquilizers, and time on the psychoanalytic couch, being a good mother meant resisting the robot nurse, embracing the bassinet, and devoting oneself full-time to bringing up baby."[15] It was a tension that foreshadowed the one that would come to the classroom: being a good school meant buying educational machinery, but being a good teacher meant resisting roboticization.

One of the more horrified letters to Skinner began "Dear Sir, Your Artical [sic] in Better Homes and Gardens was awful. No one with normal sense would do such a thing. Think of the wives of soldiers and sailors. They have children to take care of, sometimes their husbands are dead—do they put their baby in a contraption—No! They give their babys [sic] the love and affection they need. If you don't care what happens to your baby why have one?"[16] "A Reader of the Times" wrote to the District Attorney of Bloomington, Indiana (where Skinner was, at the time, a faculty member at the university), "I have read about this professor who thinks he can rear his little child by depriving her of social life, sun and fresh air. Can't you people of the law do something about this. . . . These crack-pot scientists. . . . It is the most ridiculous, crazy invention I ever heard of. Caging this baby up like an animal, just to relieve the Mother of a little more work."[17]

The appeal of Skinner's crib did find its way to high-profile households. Leila Roosevelt Denis, the wife of filmmaker Armand Denis (and daughter of a cousin of Theodore Roosevelt), wrote to Skinner asking for instructions about building a baby box, suitable not only for a human baby but also for the chimpanzee babies the couple had in their care at the Anthropoid Ape Research Foundation in Florida.[18]

Some twenty years after the publication of the *Ladies Home Journal* article, Eunice K. Shriver, wife of Sargent Shriver (and sister of President John F. and Senators Robert F. and Ted Kennedy), wrote to Skinner asking how she could get ahold of a crib and wanting more details on "the cost, size, etc."[19]

Skinner also made sure that his colleagues were aware of the air crib. He wrote to Columbia University psychology Edward Thorndike, who'd expressed interest in purchasing one for a new grandchild.[20] He wrote to the author of Stanford-Binet intelligence test, Lewis Terman, suggesting he run an experiment with twins—one raised in an air crib and one in the "usual way." "That seems to me the quickest way to get a fair comparison with respect to genetic and environmental influences."[21] Skinner remarked to Terman that "one or two of the psychiatrists or pediatricians who have lately gone in for 'mother love' as a panacea have raised objections, but these are founded upon ignorance as to the device in actual use"—a reference perhaps to the publication in 1946 of the wildly popular parenting book authored by Dr. Benjamin Spock, *The Common Sense Book of Baby and Child Care*—a book that was diametrically opposed to this behaviorist, technological model of childrearing and that instead encouraged parents to be more affectionate and more "natural."

Skinner received a sufficient number of encouraging letters about the crib—letters asking for instructions on how to build something similar—that he decided to try to turn the idea into a business. Shortly after the publication of the article in *Ladies Home Journal*, he received a handwritten note on letterhead that read "Display Associates" from a man named J. Weston Judd. Judd suggested the name "Heir Conditioner" and revealing that his one month old son "has a damp, cold and dirty Cleveland winter facing him," offered to "manufacture a few on a special order basis in order to test the market."[22] Skinner responded that the proposition "appeals to me," but said that he'd contacted several attorneys who had told him they didn't believe that the crib—"merely an air-conditioned room no matter what the special size"—was patentable.[23] That being said, Skinner added, "a registered trademark and an early appearance in the field could do a lot." Outlining his hopes for advancing child care as well as early childhood development research (and admitting "if there is any money to be made from the idea, I am not averse to getting a reasonable share of it"), Skinner asked for a few more details about what Judd had in mind for the manufacturing before proceeding. But Skinner indicated that, regardless, he was very amenable to a business partnership with Judd—despite his being a total stranger.

Judd responded, this time with a typewritten letter (still on Display Associates letterhead), proposing to form a company called "The Heir Conditioner Company" that would be the sole proprietor of all the copyrights, trademarks, and patents and that would grant Skinner a quarterly royalty payment of 5 percent on each unit sold. The plan was to make a simple model costing less than $100 with a deluxe model perhaps costing $200 (about $1,500 and $3,000 respectively in today's dollars). In exchange, Skinner would help supervise the research, development, and testing of the cribs and turn over any ideas for improving them. While supplies were still in short supply because of the war, Judd was confident that he could get what he needed. "We have a small shop set up primarily to make displays and display fixtures,"

Judd informed Skinner. "We have all the power woodworking tools, spraying and finishing room equipment necessary to produce the 'Heir Conditioner' in lots of fifty at a time. To handle more than that we would require additional space—however I would like to have more problems of that nature."[24]

The two men continued to correspond frequently, with Judd asking for clarifications about some of the design elements and Skinner offering suggestions for improving the crib's manufacture and forwarding inquiries that he'd received about purchasing one. In December 1945, formal paperwork to incorporate the Heir Conditioner company was drawn up, with Skinner to retain 12 percent of the common stock for $500, plus a credit of $1,000 for assignment of the invention; with J. Weston Judd retaining 25 percent for a credit of $3,000 for expenses he'd incurred; with Julian Bobbs, the retired president of the Bobbs-Merrill Publishing Company, retaining 50 percent for a $6,000 cash investment; and with a fourth investor still to be named.[25] (Skinner had tried but failed to raise the additional cash from his father. "He is naturally a cautious man," he apologized to Judd, "and has never had any faith in my business judgment."[26]) But as the end of January 1946 approached, Skinner and Bobbs realized it had been weeks since either had heard from Weston, and the incorporation documents remained unsigned. The two became increasingly concerned—Bobbs about his investment; Skinner about his reputation.

Skinner, who'd promised his neighbors one of the first commercial models, had to inform them in February that nothing was ready. He continued to write to Judd, who became less and less responsive. When Skinner's neighbors did finally receive a crib, the thermostat didn't work, and it smelled strongly of paint. "I have lost a lot of sleep trying to see a way out of the whole mess," Skinner wrote to Judd in April. "I can't see that any progress has been made during the past six months. On the contrary a lot of time, goodwill, and publicity has been lost."[27] After traveling to Cleveland himself to investigate Judd's disappearance, Skinner had to write to customers—those who'd paid in advance and those who thought they were on a waiting list—and confess that there were no cribs to be had. "As you must realize from your dealings with Mr. Judd," he wrote to Mrs. H. M. Gibson of Memphis, Tennessee, "he proved to be incompetent and unreliable and has now withdrawn from the picture. His former partners and I are undertaking to put things in order, and I should like to make a personal request for your sympathetic cooperation."[28] Gibson reported Judd to the Cleveland Better Business Bureau, as she had paid $200 for a crib—about $2,500 in today's dollars—but it had never been delivered.[29] Another potential customer threatened to write to the *Ladies Home Journal* if Skinner did not reimburse him. Skinner sent him a check for fifty dollars.

It wasn't the end of the air crib. But it did mark the end of Skinner's complete naiveté about business. When it came time to commercialize his teaching machine a decade later, he had learned his lesson. Or at least, he was confident that this time, he would be able to navigate the legal and financial wrangling much more effectively, with less damage to his pocketbook and his reputation. He was, after all, now a Harvard professor. He had better resources. He had a good lawyer.

In June 1956, Francis Keppel, the dean of Harvard's Graduate School of Education who'd made the first introductions between Skinner and IBM, inquired how the work was proceeding. Skinner replied that negotiations were still "midstream" and admitted, "I am not yet certain that we can work out anything which will be mutually satisfactory. I am not convinced that IBM takes the matter as seriously as I do."[30]

Keppel had encouraged Skinner to apply for a grant from the Ford Fund for the Advancement of Education to pursue this research, and the foundation had awarded him $25,000 (approximately $240,000 in today's dollars) to build new teaching machines for experimental use at Harvard— "mechanical devices adapted to college teaching in elementary language and science" that permitted "immediate grading of multiple choice answers" and required the student to "compose answers, thus demanding creative activity rather than mere recognition."[31] Skinner received a number of other, small grants—enough to hire Lloyd Homme, "an imaginative, if undisciplined young psychologist who could take a year's leave from the University of Pittsburgh,"[32] as well as Susan Meyer, a graduate student from the University of Buffalo who was tasked with writing the arithmetic program for IBM.

IBM and Skinner finally signed a contract in November 1956.

In January 1957, Skinner wrote to IBM with a number of outstanding questions: had IBM completed the necessary paperwork to have Susan Meyer paid as a consultant? Had IBM thought about a name for the machine that could be trademarked? (Skinner said he very much liked "Autostructor.")

Would IBM like to look at the machines he'd built with the Ford Foundation grant money? "If IBM is not interested," he told the company's leadership, "I would probably approach a large publishing house, such as McGraw-Hill, who have been interested in various instructional devices."[33] Skinner did show these prototypes to IBM, but company executives felt it would be better for IBM to put all its effort into the teaching machine it was already developing.

In April, Skinner again wrote to IBM, concerned that the project was not keeping to its proposed schedule. "The time is near when I should make some definite arrangements for working in a school next fall," he reminded his contact at IBM, Les Bechtel. "Have you any word from the engineers as to progress on the machine? Do they feel that they will have machines available by September?"[34] He suggested that, if the company wasn't ready for testing at a school, Susan Meyer might still be able to work with individual students. "This would give us at least a preliminary run on its suitability." He wrote again in June: "Have you any word on the availability of machines for a test this fall? I really should attempt to make arrangements before the summer holidays begin."[35] A few weeks later, George Youngdale, Jr. responded: "Les Bechtel has been promoted to Assistant Sales Manager of the Division and no longer has the responsibility of the Product Planning area."[36] Youngdale said that he was taking over Bechtel's responsibilities but was "unable to offer any incouragement [sic] on the availability of the machines for the test you mentioned, as based on present circumstances, it does not seem probable that they would be available."

Skinner continued to send letters to IBM, writing in August to inform the company that he planned to display his "write-in" machines at the American Psychological Association conference. *LIFE Magazine* also planned to do a story on teaching machines. Did IBM want any acknowledgment? "In view of the fact that we have not completed your teaching machine as yet," Youngdale replied, "we would request that IBM's name not be mentioned" to the reporter. Furthermore, Youngdale said, the IBM machines would not be available in September as initially planned.[37] It would be February 1958 before they could be installed in schools for testing. Soon after, that target date too became "unlikely."[38]

Something happened in the fall of 1957 that changed the perceptions—and perhaps the priorities—of education and educational technology. On October 4, the Soviet Union successfully launched Sputnik, the world's first artificial satellite, into space. The event thrust education to the forefront of the minds of both the popular press and politicians in Washington, DC. "The launch of the Soviet Sputnik satellite in early October 1957 intensified the Cold War-related push for technological advancement through educational reform," Wayne Urban chronicles in his book on the National Defense Education Act, which was passed the following year.[39]

"For Dwight D. Eisenhower," Urban argues, "Sputnik presented a challenge that was much more political than it was military or economic." The president was confident in the intelligence information that assured him that, even though the Soviets had beaten his country into space, the United States retained its scientific and technological superiority. Eisenhower was not impressed by "the Soviet Fear," and he had to reassure the populace to that end.[40] In order to assuage fears about the Soviets' successful satellite launch, Eisenhower promoted "a short-term infusion of money and energy into the nation's schools to assure its citizenry that corrective steps were being taken. That infusion was advocated most prominently for the areas of science and mathematics education by a group of the president's scientific advisors."[41] The Science Advisory Committee argued that the United States was *not* behind the Soviets scientifically, but that the Soviets emphasized science in their schools more than Americans did, and that this emphasis, unless countered, would lead to Soviet scientific superiority within a decade. The success of Sputnik became a story of America's educational deficiency.

The school system had become too lax, many critics charged—a result of the adoption of certain aspects of "progressive education," something that was often linked to the work of John Dewey (even when contrary to the kinds of practices the philosopher had actually advocated). Dewey had argued that student interest and curiosity were the key to learning, and many mid-century educators had endorsed his ideas, promoting project-based and problem-oriented studies, for example, and introducing new subject areas that were supposed to be more culturally and socially relevant. One of the most outspoken critics of progressive education was US Navy Admiral Hyman Rickover, who published a book in 1959, *Education and Freedom*, which held the education system responsible for a decline in American society, charging that the country had become intellectually and physically "soft" due to the lack of rigor in school discipline and curriculum.[42] The launch of the Soviet satellite might have triggered a push for more science education, but many newspapers and magazines echoed the criticisms of Rickover and others and called for even broader educational reforms, suggesting that there was a direct link between the country's failure to put a satellite in space and its schools' embrace of progressivism.

An article in *Barron's* in 1960 exemplified how the press treated the launch of the Soviet satellite and the threat of a Soviet scientific supremacy, making the case for more technological investment in US schools:

> At present, the U.S. is spending an estimated $15 billion annually to prepare 42 million children for the rigors of life in the space age. Many Americans, however, are convinced that this effort is woefully inadequate. While Russian schools, according to reports, are grinding out scientists and technicians like sausages, the vast American educational plant is so overcrowded in many areas that children only attend on a part-time basis and nationally there is a reported shortage of 200,000 teachers. In response to the huge and cry to "do something," what amounts to a new industry, devoted to the manufacture and sale of educational devices, is arising. It makes thousands of gadgets for school laboratories, turns out school furniture, puts lessons on film and tape, supplies components for educational television systems and develops such specialized units as the "electronic learning center."[43]

Sputnik was, if nothing else, a good marketing pitch for technological change. As one CEO of a scientific equipment company told *Barron's*, "I have one of the world's greatest salesmen working for me, and I don't pay him a cent of commission. His name is Khrushchev."[44]

Although Sputnik did alter the national narrative about the quality of schools and did lead to more money for educational endeavors that aimed to boost student achievement in math, science, and foreign languages, IBM showed little sign it planned to respond to the event by expediting its work with Skinner, even though math, science, and foreign languages were precisely what the professor's machines purported to teach. IBM seemed to feel no pressure to react to Sputnik specifically or to the more generalized crisis that was supposedly brewing across American classrooms. As Skinner's biographer, Daniel Bjork argues, "the inventor and the company" were "working on different timetables. Skinner felt an urgency to market his teaching machine that IBM never shared. Moreover, IBM never properly acknowledged his role as the inventor."[45] Skinner remarked to a friend who was trying to get a different company interested in developing his machine that "'all [IBM] need[s] me for (and all they will pay me for) is the construction of material to be used in the machines. Nothing for the idea, nothing for testing models, nothing for prestige in the field—in short nothing for my good will. I had heard this of IBM but never quite understood.'"[46]

Perhaps because of the delays in making progress with IBM, Skinner had started to change his mind that one company should produce both the teaching machines and the programs. In March 1958, Skinner had lunch with William Jovanovich, the president of Harcourt, Brace, and Company, hoping to convince the publisher that there were opportunities to produce teaching machine materials for the high school and college level. Skinner confessed to Jovanovich that he was frustrated with IBM and was looking at other machine manufacturers—perhaps Comptometer or American Voting Machine or Lockheed Aircraft—that might be more serious about building his device. Jovanovich seemed intrigued with the idea, and he wrote back to Skinner in April, indicating that "Harcourt Brace is prepared to publish the programming materials for use in these machines."[47] He wrote again the next day, saying that he'd had a long talk with R. E. Zenner, the vice president of Union Thermo-Electric Corporation, a subsidiary of Comptometer, who was impressed that Harcourt Brace was willing "to throw itself into this project so promptly without contractual terms."[48] Comptometer might be willing to manufacture a machine for use with Harcourt Brace materials, Jovanovich speculated. "It helps to know that someone is willing to make the blades for the razor," Zenner had told him. It was an opportunity Skinner was keen to be a part of, particularly in light of the problems he was experiencing with IBM.

Indeed, around the same time, IBM had pushed back its target date for the new model again, this time to April. When April rolled around, the model still had problems. Skinner urged executives at Harcourt Brace to reach out directly to IBM CEO Thomas Watson. "I think a little pushing at a higher level would help all of us," Skinner wrote to the publisher's treasurer, John McCallum. "As I indicated to you IBM has been only moderately warm on the teaching machine until very recently. Now several engineers have caught fire, but things are still going slowly. It would do no harm for Watson or someone close to him to be reminded that they have a teaching machine project."[49] After meeting with IBM, McCallum drafted a memo for his employer, in order to help the publisher decide if it was worthwhile pursuing a deal with IBM. "At the moment," McCallum wrote,

> IBM does not seem to have a very clear idea as to just when the experimental model of the machine will be completed or how they will go about experimenting with it once it is completed. They have been planning, I judge, to have their own salesmen sell the eventual machine just as their typewriters and other business machines are sold. They seemed to be willing to accept my contention that the machine manufacturer could not successfully sell the machine in schools, and that the manufacturer could not successfully prepare tapes—that both these activities could better be performed by a textbook publisher.[50]

Harcourt Brace was prepared to insert itself into the teaching machine business.

McCallum wrote to Skinner and shared the memo with him. He told Skinner that IBM's executives "recognize that they have been seriously at fault in not getting ahead more promptly with your machine. I think the delay stems primarily from their inability, thus far at least, to evaluate the market. It seems pretty clear to me, that is, that if they can't count on a sale of a quarter of a million machines a year they will not want to go into this business."[51] That being said, McCallum added, if IBM did move ahead, the company would make other machines and "all sorts of complications will arise." McCallum regretted, he admitted, not working with IBM more closely all along. In the interim, McCallum had still been trying to strike a deal with Comptometer, whose slowness in working toward any deal "may prove to be a blessing—if, by chance, we find IBM ready to do business."[52]

Skinner had hoped that the publisher would be better positioned to advocate for the teaching machine's marketability. Surely dealing with a publisher—something Skinner had experience with as an author—would be easier than dealing with IBM. Instead, what he found was added complexity as Harcourt Brace had many possible schemes and connections it wanted to pursue.

Harcourt Brace had also made overtures to Bell Labs, the research laboratory of the Bell Telephone Company. Skinner had earlier recommended Thomas Gilbert, a young psychology professor at the University of Georgia, to Bell Labs when the company expressed an interest in researching teaching machines. Gilbert had built his own devices—"at first a very complicated one, and now a quite simple one . . . manually operated, and so put together that it can use ordinary mimeograph material on 8½" x 11" sheets."[53] Bell was manufacturing around a dozen of the machines, and Gilbert said he'd sell McCallum one for $10 or $15. Gilbert didn't think the company planned to make them as a commercial product; and even if Bell decided it would use the machines with trainees, it would only need about ten thousand devices, McCallum estimated—a figure that was still far from the quarter of a million machines he estimated IBM would consider the minimum sales threshold to entering the market.

As Skinner's relationship with Harcourt Brace deepened, he realized he should check to see if anything he was doing might violate the contract he already had in place with IBM. His attorney confirmed in August that that contract "excludes testing machines, multiple choice teaching machines and other teaching machine in which a composed or written answer is compared by the student with a printed or otherwise uncoded answer subsequently presented in the machine"—precisely the type of machine he was trying to develop with Harcourt Brace.[54] Skinner was safe, contractually at least, to proceed. Harcourt Brace asked its attorneys for clarification too about what Skinner could and could not do, legally, with teaching machines. "IBM has, according to Skinner, been both lackadaisical and confused in their handling of the machine and the teaching materials to go with it. Little has been done—in fact, the machine prototype, as we understand it, has not been finally made,"[55] Jovanovich wrote, admitting that Harcourt Brace's interest in Skinner had, perhaps, reawakened IBM's. Harcourt Brace's attorney responded, as McCallum informed Skinner, that patents needed to be secured for the two machines Skinner had developed with the Ford Foundation money.[56] The patent law firm of Kenyon and Kenyon, McCallum said, would take care of this, and he urged Skinner to speak to an attorney before meeting with IBM at the end of August.

In September, McCallum wrote to Skinner that it was time to work out "an interim agreement"[57] between the professor and the publisher. Skinner would agree to act as a consultant for Harcourt Brace, advising the company on "the development (should there be any further developments), the manufacture, and the distribution of the machines themselves" as well as on securing authors for teaching machine materials. Harcourt Brace would pay Skinner $6,000 for the two-year duration of the agreement—about $50,000 in today's dollars—and Skinner would assign the forthcoming patents for Machines Number 2 and 3 to the publisher.

In October, R. E. Zenner, the vice president of Union Thermoelectric Corporation, wrote to Skinner announcing that its parent company, Comptometer, was ready with the first model of its teaching machines.[58] But Zenner seemed to have overpromised, because when Skinner and McCallum met with the company for a demo, they were, in Skinner's words "still a long way away."[59] Skinner had no confidence that Comptometer would be able to build a working teaching machine. Nor did McCallum, who wrote to the head of Comptometer and told him that the October visit with the company was "a great disappointment. The model you should use failed completely to fulfill the desired objective and was entirely inadequate and unacceptable."[60] Under these circumstances, Harcourt Brace felt it could not continue with even an informal plan to have Comptometer develop a teaching machine for the publisher's planned programming materials.

The IBM model was still not functioning correctly either, having been recalled to the Kentucky manufacturing location for a safety inspection in September. One working model was given to Skinner, and Susan Meyer was able to use it with a few children in a school in Somerville, Massachusetts. "The evidence was clear: The children liked the machine and learned quickly," Skinner later wrote in his autobiography.[61] But with just the one machine, there was no way to run experiments that would test the efficacy of programmed instruction on a whole class.

As the companies that he'd signed contracts with all seemed to stumble, Skinner became increasingly agitated about what other competitors were doing. In November, Skinner wrote to McCallum to complain: "I don't like what is going at the Bell Telephone Labs, and wonder whether you might ask Angulo [an attorney for Harcourt Brace] about it."[62] Skinner reminded McCallum of how he had introduced Tom Gilbert to Bell Labs, who had "sent out a general notice to engineers or some other group of employees announcing the machine age in education, and pointing out that Bell Telephone Labs were in the vanguard of this development." Skinner had also learned that Bell had hired John Gilpin, who had briefly worked in Skinner's teaching machine lab at Harvard. Gilpin had not only taken a copy of Susan Meyer's arithmetic program back to Bell's research facility in Murray Hill, New Jersey, he had offered her a job "behind my back," Skinner wrote with disgust. "The tone of his report of activities at the Bell Labs suggests that they are making an all out effort to invent the teaching machine," he continued, offended that anyone would dare write him out of the history. Skinner urged McCallum to have the publisher's attorney look into a possible infringement claim, but McCallum responded coolly, reassuring Skinner not to worry about Bell Labs or any legal matters.

While Skinner fussed about Comptometer and raged about Bell Labs, his relationship continued to unravel with IBM. In November, IBM's George Youngdale wrote to inform Skinner that the company was poised to make its final recommendation for what to do about teaching machines. It had been five years since Skinner had come up with the idea for his teaching machine while visiting his daughter's class. There was still no "Skinner" machine on the market. And by the end of the year, IBM had made its decision, opting to terminate its agreement with Skinner and assigning his teaching machine patent back to him.[63]

In January 1959, Skinner wrote to H. W. Miller, IBM's vice president, to inquire what would happen to the "rather extensive materials" that were designed by Susan Meyer in order to test the machine. Skinner said he recognized that since Meyer had been paid by IBM as a consultant that the company might lay claim to the tests.[64] Instead IBM signed over all of Meyer's work ("rights, title and interest in the results"[65]) to B. F. Skinner—to the professor, that is, not to the graduate student who had actually performed the labor.

"Hold everything," an IBM engineer named Hal Robson wrote to Skinner in March of 1959. "The decision to drop the machine had been made by the director of the Typewriter Division, who preferred to use any available money to keep the IBM typewriter ahead of its competition," Skinner later wrote in his autobiography, "but the Data Processing Division might take over. An effort would be made to have the young children of Thomas J. Watson, Jr., President of IBM, try the machine."[66] But the renewed effort with IBM failed too. The company had already embarked on its development of computers and was building a device that would combine a typewriter and a computer to teach math—the IBM 650.

Meanwhile, Harcourt Brace had decided not to pursue its teaching machine plans, and Skinner's agreement with the publisher was terminated in March 1959. So, Skinner began to look elsewhere for someone to manufacture his machines—McGraw-Hill, RCA, Hughes Aircraft, Western Design, the Monroe Calculator Company, and Foringer, for starters. These companies all displayed strong levels of interest at first, but for various reasons they were not able to meet with Skinner's approval.

Skinner wrote to IBM's Robson in June that "the most promising opportunity at the moment seems to be the Rheem [Manufacturing] Company which is really planning a fine program including money for public relations to woo the teacher organizations." He hoped that Rheem would woo Robson away from IBM too. "My only objection," Skinner added, "is that they do not at the moment have a plant ready to put into operation on prototype models although they have plenty of money for setting one up."[67]

After spending the spring and early parts of the summer of 1959 negotiating with various companies to no avail, he expressed much the same thing to his attorney, Donald Rivkin: "I want to place an all-or-nothing bet on Rheem. If they want out at the end of eighteen months, then I will no longer have any part in the production of teaching machines."[68]

5

B. F. SKINNER TRIES AGAIN

Rheem Manufacturing Company was founded in 1925 by two sons of William S. Rheem, the president of the Standard Oil Company. The company initially manufactured galvanized steel drums for the petroleum industry, expanding in the 1930s and 1940s to produce other kinds of steel containers as well as water heaters, and turning during World War II, as all manufacturers did, to supplying the war effort—in Rheem's case, building parts for aircrafts. By building teaching machines, Skinner later wrote in his autobiography, Rheem was hoping to diversify, and the company painted for Skinner "a glowing picture of the financial rewards which lay ahead."[1]

Before finalizing a deal with Rheem, Skinner had informed his contact at the company, Dean Luxton, that he was pursuing potential business relationships with other manufacturers and that Foringer & Company in particular seemed well positioned—even better prepared than Rheem—to develop the prototypes. In response to the possible competition, Rheem appeared to move quickly, locating a shop that could build a teaching machine model based on Skinner's designs and drawings—an improved version of the "disk" machine that Skinner had been using in one of his courses at Harvard. Luxton sent Skinner a draft of a proposed agreement in late June 1959, "the terms of which were much less opulent than those we had first discussed," Skinner later admitted.[2]

Even before the contract was finalized, Skinner was off and running with this new relationship. He wrote a letter to Luxton in July that contained several suggestions for how the devices should be displayed at the APA meeting, as well as how Rheem might name them: "the most suitable Greek root seems to be didak. European languages in general use something equivalent to autodidak to refer to self education. The word Didak is brief, possibly a little too much like Kodak, but otherwise quite distinctive." (Skinner did recognize that teachers might not like the connotations of self-instruction too much.) Skinner also suggested numbering the machines—the Series 100, 200, 300, and so on—noting that this was IBM's strategy, although it used its own company name rather than composing a new brand name for each product. "Possibly you will want to let the question of the name ride for a while, although I think it is desirable to register Didak and Autodidak, even if they are not eventually used. There aren't very many possibilities. Other terms I have thought of without much favor are 'Instructomat,' 'Autostructor,' or possibly just 'Structor.' End of communication in RE: Names of Teaching Machines," Skinner closed the two-page missive.[3]

But it's the opening sentence of this letter that perhaps best demonstrates what it was like for any company to work with the Harvard psychologist: "I am afraid you are going to get a chaotic assortment of letters from me, but it seems easiest to write on separate points as they come up, even if it means sending you two or more letters a day." The following day, Skinner wrote to Luxton four times. He wrote him three more times that week, in addition to sending a letter to a Rheem engineer, two to his attorney, and several to colleagues at Harvard asking their opinion on whether or not Rheem using Skinner's name in their materials violated any professional ethics clause. (Stanley Smith Stevens—"Smitty"—responded to Skinner, "I would go ahead as you have proposed."[4] But the chairman of the Harvard Department of Psychology, Edwin B. Newman, was a bit more cautious: "If the machine turns out to be no good and it has been designed in consultation with you, you are the person to suffer. I think your strongest protection here is that the advertising must not overstate your contribution to the ultimate product."[5]) By the end of the month, while vacationing in Maine, Skinner penned a note to Luxton stating with some satisfaction that everything appeared in order.

On August 3, Rheem Manufacturing Company sent over the final contract agreement, and Skinner's attorney wired him a few days later, urging him to sign. The agreement was settled on August 13.[6] Skinner's third attempt to commercialize his teaching machine concept was officially underway.

Correspondence to Skinner continued to pour in, particularly after the publication of his article on teaching machines appeared in *Science* in 1958. He quickly began forwarding the names of interested buyers to Rheem, hoping to show the company that there was indeed a sizable demand for the devices. Seeing one of the key elements in his role as marketing and promotion, Skinner made a number of plans for the fall of 1959 to travel and speak about teaching machines, writing to Rheem encouraging the company to consider sponsoring some of these conferences.[7]

But Skinner's attorney Donald Rivkin was worried, because almost as soon as the contract had been signed, Rheem indicated it wanted to modify the agreement, adding a clause that would give it the right to terminate at the end of the year. Following Rivkin's suggestion, Skinner wrote to Rheem saying that no alternation to the original agreement was necessary.[8] Yet.

"Teacher Machine to Be Ready in '61," the *New York Times* reported in September 1959 from the APA Conference in Cincinnati.[9] "The machines will be in mass production next year," the paper announced, relaying the news about teaching machines from the psychologists' annual conference and announcing the deal the Harvard professor had struck with Rheem. "The only hitch at the present, they say, is the lack of teaching programs designed with the machine in mind."

It was not the only hitch.

Skinner sent a memo to Rheem in September, detailing a number of problems with the Didak 101, the "pre-verbal" machine designed to teaching spelling to young children. It was too noisy. Its lid was too easy to unlatch, and too many sensitive parts could be reached by a child who managed to open it. Skinner was not convinced that new features—green lights and a timer—were "worth the added expense."[10] Rheem had added these to reward the child, Skinner recognized, but he argued that simply getting the answer right and moving on to the next frame was positive reinforcement enough. Skinner sent a separate memo the same day, suggesting detailed improvements for the 501 model, the device based on the one he'd been using with his students at Harvard and the prototype of which he'd displayed at the APA meeting.[11]

Despite his falling out with Harcourt Brace, Skinner still believed that the publication of high-quality teaching machine *programs* was necessary for the success of teaching *machines*. Skinner wrote to Luxton mid-September, asking for him to clarify whether or not Rheem had any interest in developing the programming materials or if he was free to approach publishers to that end. Skinner hoped that Rheem did not want to pursue designing the programs themselves; nor did he wish to see the company sign an exclusive agreement with just any program maker. "Publishers are in an unusually good position to evaluate the market, to arrange for publication, and to turn the distribution of materials over to their well established sales forces," Skinner cautioned. "It seems to me that it would be very difficult to develop an adequate distributing organization which could compete with a well established publisher."[12]

In addition to his interest in Rheem's plans, Skinner remained attuned to what his previous business partner Harcourt Brace was up to. "Harcourt Brace is apparently pulling a quite unethical trick," he complained to Luxton.[13] Noting that "it was clearly understood" that when Harcourt Brace dissolved its agreement with Skinner that the publisher intended to get out of the teaching machine business, he had learned that it had recently formed a new group that was continuing that work and using a machine that looked very similar to the one Skinner had patented. He told Luxton he'd asked his attorney to look into the matter and ascertain if this was a case of patent infringement. Regardless, it was "clearly a breach of good faith," he insisted.

Skinner recognized that, despite his stature and his role in inventing and promoting teaching machines, he had no monopoly on the idea. Other universities' psychology departments were pursuing teaching machines as research *and* as commercial endeavors. "I think we must anticipate that someone, Harcourt Brace or perhaps the Hamilton College group, are going to come out with a film machine, probably heavily overdesigned and expensive compared with our 501," Skinner continued in his letter to Luxton.

Skinner's attorney, Donald Rivkin, responded promptly to Skinner's concerns about Harcourt Brace's renewed interest in the teaching machine business, crafting a letter for Skinner to send to John McCallum.[14] But Rivkin seemed just as concerned about Skinner's dealings with Rheem as he was about Skinner's previous business partner. "Rheem's continued refusal to comment upon or even to acknowledge your various letters to them puzzles, and to some extent, troubles me," Rivkin confided. He urged Skinner to put everything in writing with Rheem and to "have a certain amount of healthy suspicion concerning the people with whom you are treating." Perhaps his attorney's advice helps to account for Skinner's incessant communications to Rheem's various officers.

Things seemed better at Rheem—to Skinner at least—by the middle of October 1959. He wrote to Rivkin that the company's executives all appeared to be eager to move forward with manufacturing the teaching machines. "Interest is high," Skinner reported, "and the company proposes to accelerate the program considerably. A deal is underway to acquire a company now engaged in the distribution of school materials such as audio-visual aids, language laboratories, and so on in order to facilitate the ultimate commercial distribution of the teaching machines."[15] Rheem paid $1 million for a majority stake in the Califone Corporation that very month.[16] In the hopes of better explaining his ideas to key decision-makers at Rheem, particularly in light of a shake-up in management following the Califone acquisition, Skinner sent the president of Rheem copies of his books *Verbal Behavior* and *Science and Human Behavior*.[17]

The academic literature on behaviorism did little, it seemed, to convince Rheem of the business case for teaching machines. In early November, Luxton wrote to Skinner, giving him a brief update on the company's "thinking in that area." Initially, Luxton said, the company was planning on building thirty-seven machines to give away to various educational institutions, although the company had realized that that might not be enough to cover all the "educational, industrial and military pilot classroom situations" necessary to ascertain how best to proceed with entering the market. "Now with the formation of Rheem-Califone a new management is being formed which will have jurisdiction over the number of machines produced, the nature in which they will be sold, and the conditions of their placement," Luxton wrote.[18] He could no longer even promise thirty-seven machines, although he assured Skinner he was "relatively certain" that *some* classroom models would be produced. However, "no specific information can be transmitted to you until the new management is formed," he said.

The work on Skinner's teaching machine was, once again, at a standstill.

Two days later, Skinner received a form letter from the Rheem Califone Corporation, thanking him for his interest in the company's line of "automated learning instruments" and promising him he was on the mailing list for when more information became available.[19] Rheem was not paying attention to *any* of the details, neither in manufacturing nor in marketing his machine, Skinner suspected.

Skinner met with directors of Rheem in December, and as he feared, the changes he'd suggested for the teaching machines under development had not been made. Indeed, the machines he was shown at the December meeting were identical to the prototypes he'd reviewed in September. "Nothing has been done to develop it in spite of the fact that literally hundreds of people are waiting for models," he wrote indignantly to Rheem. "I am sure you are as unhappy about a decision of this sort as I am. Three very valuable months have been wasted, particularly since Rheem's strength in the field will depend upon an early appearance with an inexpensive and satisfactory model."[20]

"It is true that the progress since September has been very slow," lamented Robert Metzner, a new contact at the newly reorganized Rheem Califone Corporation, adding, "but you should not feel that it has been due to a lack of cooperation." The merger had slowed all of the company's activities, but Metzner promised that new and larger Califone facilities would make the production and distribution faster and would quickly compensate for the lost time. He assured Skinner that the engineers—indeed, everyone at the company—were listening to the psychology professor's suggestions.[21]

Skinner continued to correspond with the company's engineers, discussing features and changes to features of the machines and offering feedback on—and suggesting many, many changes to—new designs.[22] He wrote to his attorney in February with renewed optimism. "Things are moving steadily at Rheem. I am less and less anxious to extend my obligations to them and may settle for something very close to the original agreement in order to keep control of my own time."[23] A new contract with Rheem was in the works, and Skinner wrote to Donald Burdorf, a Rheem engineer, in late February: "So far as I can see, the comments I have sent are all those which need to be made. I believe the machine is ready to go in essentially its present form."[24]

In March, Skinner informed I. G. Davis, Rheem's director of planning, that he no longer thought it necessary to travel around the country giving talks introducing the concept of the teaching machine. "I was glad to do this when the idea was unfamiliar and when erroneous ideas were circulating, but the general point now seems to have been put across. What groups want is specific information about machines and programs." Noting that his own work at Harvard was in jeopardy because of his promotion of teaching machines, Skinner said, "I could spend most of my time giving talks to various groups of parents, teachers, administrators, and so on, but if I were to do so my own work here would suffer."[25]

Yet Skinner also wanted to make sure he remained at the center of the growing teaching machine movement. As such, he balked at some of the restrictions that he felt Rheem was placing on him, curbing his interactions with other companies and by extension other peers who were interested in teaching machines. "I cannot retard the development of my research by refusing technical help which is offered to the project without any strings attached, simply in order to protect future activities at Rheem Califone."[26]

Skinner was far less interested in the commercial success of the teaching machines—and certainly of Rheem Califone—than he was the total transformation of the American education system. So, he decided to try to appeal to one of the best-known and most powerful people in the field, someone who had also recently captured public attention with his ideas about reshaping schools: James Bryant Conant.

Conant had started his investigation into the quality of US high schools in February of 1957, before the launch of Sputnik in October of that year, "before the appearance of the Russian satellites led the American people to take a fresh look at their educational system," as the head of the Carnegie Corporation described the project.[27] But by the time Conant's Carnegie-funded report, *The American High School Today,* was released in early 1959, the political landscape had shifted. There was a heightened sense of concern about the shape of the public school system and much more urgency to the need to address its perceived problems.

Conant's experience with both education and government service was extensive—he had been the president of Harvard University from 1933 to 1953, had served on the National Defense Research Committee and on the Committee on Scientific Aids to Learning (in both cases, with Vannevar Bush and the latter with Columbia University's Ben D. Wood), and had been appointed by President Eisenhower as the United States High Commissioner for Germany—a post he left immediately prior to pursuing research into the quality of public high schools for the Carnegie Corporation.

The question to be answered by his work, Conant wrote, was this: "Can a school at one and the same time provide a good general education for all the pupils as future citizens of a democracy, provide elective programs for the majority to develop useful skills, and educate adequately those with a talent for handling advanced academic subjects—particularly foreign languages and advanced mathematics?"[28] Conant's focus was on the "comprehensive high school"—that is a high school whose programs "correspond to the educational needs of all the youth in the community"[29]—as opposed to specialized high schools that taught students with particular aptitudes or those that offered specific vocational training.

Despite its framing as a thorough analysis of "the American high school today," Conant restricted his report to high schools in just twenty-six states and primarily to schools in cities with populations between 10,000 and 100,000, conducting on-site visits to fifty-five different schools.[30] Not only was this therefore a study of *urban* schools, but, with the exception of Texas and Virginia, the research did not include a single school in the South. As Conant later admitted, *The American High School Today* completely avoided the topic of racial segregation—an incredible oversight considering the Supreme Court decision in *Brown v. Board of Education* just a few years earlier—uttering "not a word to indicate that certain schools I visited were comprehensive only in so far as white youth were concerned."[31]

By ignoring the grotesque inequalities in education between Black and white students, regardless of their geographic location, Conant could make rather simple, albeit sweeping suggestions for what needed to happen to improve the American high school system: "I can sum up my conclusions in a few sentences. The number of small high schools must be drastically reduced through district reorganization. Aside from this important change, I believe no radical alteration in the basic pattern of American education is necessary in order to improve our public high schools." If all schools functioned as well as some of the best schools he'd visited, Conant said, things would be fine. "If the fifty-five schools I have visited," he continued, "all of which have a good reputation, are at all representative of American public high schools, I think one general criticism would be in order: The academically talented student, as a rule, is not being sufficiently challenged, does not work hard enough, and his program of academic subjects is not of sufficient range." There were also significant major differences in the kinds of classes that students were encouraged to take based on gender. "The able boys too often specialize in mathematics and science to the exclusion of foreign languages and to the neglect of English and social studies," he wrote. "The able girls, on the other hand, too often avoid mathematics and science as well as the foreign languages."[32]

"Perfectionists will complain that the Conant Plan is no ticket to Utopia," wrote the *New York Times* education columnist Fred Hechinger in his review of Conant's book. "This is its strength. It offers a 'do it yourself' reform so practical that the school superintendent who says it can't be done signs a public confession of incompetence."[33] "The Conant report seems almost certain to make educational history," the journal *Hispania*, published by the American Association of Teachers of Spanish and Portuguese, effused in its review, calling it "one of the most significant books on education ever to be published in our country."[34] The journal was, no doubt, keen on Conant's recommendations for improving the high school system as they included an expansion of the study of foreign languages, beyond the two years schools typically offered.

The American high school had seen significant changes over the course of its existence, and the expectations of *who* the high school should serve and *what* it should teach had evolved as well. Sputnik simply heightened the public's sense that it wasn't doing any of this well. As David Tyack and Larry Cuban write in their history of school reform *Tinkering Toward Utopia*, in 1900 just one out of every ten youth age fourteen to seventeen was enrolled in high school; by 1940, seven in ten were. The number of students that graduated from high schools also rose sharply in these years: just 8 percent of students finished high school in 1900. By 1940, 51 percent did, by 1960, the figure had increased to 69 percent.[35] With these statistics in mind, it should go without saying that high schools in the late 1800s and early 1900s were *not* preparing students for work in factories, despite the pervasiveness of that "factory model" narrative.[36] Rather, most high schools were serving academically talented students whose parents could afford to have their child pursue secondary education—that is, they could subsist without their teenager working. As Conant wrote, "These changes could easily have been predicted in 1900 by a student of American education. He would have seen how enormous was the power of the twin ideals of equality of opportunity and equality of status; it was evident that the American people had come to believe that more education provided the means by which these ideals were to be realized."[37]

The pressure was on the school system—at all levels—to enroll still more students. Conant urged communities to build more large high schools in response and to consolidate and close the small ones, arguing that this reorganization would help address the shortage of teachers as well. "Within the next ten years, the number of Americans attending schools and colleges is expected to rise from forty-two million to sixty-two million," one journalist, Spencer Klaw, forecasted. But rather than calling for more teachers to be trained and hired, Klaw argued the solution should be technological: "Two years ago some Wall Street brokers already were predicting that as many as fifteen to twenty million teaching machines might be needed to serve them."[38] Automation was necessary to many onlookers in the mid-twentieth century—not as a replacement for teachers, but as an enhancement. As Skinner put it in his article in *Science*, "Will machines replace teachers? On the contrary, they are capital equipment to be used by teachers to save time and labor. In assigning certain mechanizable functions to machines, the teacher emerges in his proper role as an indispensable human being. He may teach more students than heretofore—this is probably inevitable if the world-wide demand for education is to be satisfied—but he will do so in fewer hours and with fewer burdensome chores. In return for his greater productivity he can ask society to improve his economic condition."[39]

Skinner was not convinced that Conant's recommendations were going to address any of the problems that the school system faced. Nevertheless, on the heels of the publication of *The American High School Today*, he wrote to Conant, asking if his former boss from Harvard was interested in learning more about teaching machines. Conant brushed the idea aside, informing Skinner that "it is certainly not worth your while to make a special trip for the purpose of instructing us."[40] Skinner was persistent and wrote back toward the end of December 1959 to inform Conant that he and his wife would be in New York for New Year's Eve and could stop by Conant's offices in Midtown in order to "discuss the issue of teaching machines with your staff. . . . I believe the issue goes somewhat beyond the machines themselves. The question really is this: can we make any use of recent contributions to our understanding of the learning process in redesigning educational practices?"[41] The two men agreed to have lunch.

Before their meeting, Skinner sent Conant a copy of the first chapter of a manuscript he was writing on teaching machines and programmed instruction: "Parents, employers, the military—these are the disaffected and they are complaining of the products of education. They demand better schools and more skillful teachers, they want students to work harder and learn more about subjects which bear more directly upon their place in the world. But it is not enough to attract better teachers by raising salaries and improving working conditions, to regroup students according to ability, to make curricula more or less specialized or more or less professional, to build more and better schools and so on."[42] It was not the wisest of writing samples to send to Conant as these were, in fact, many of the recommendations of Conant's report—recommendations that Skinner was clearly suggesting were insufficient to improve education. "We need a careful study of teaching," his manuscript continued. "No enterprise can hope to improve itself without a close look at its own technology. We should not expect much help from the past. Theories of education have never been technically productive."

In hindsight, Skinner admitted, "It was not an argument likely to appeal to a former college professor and president, for whom, according to hallowed tradition, the classroom was the teacher's castle, never to be invaded." And he recollected that "Conant opened our meeting by handing the chapter back to me with the comment: 'This is pretty shrill.' The meeting continued in more or less the same spirit and when it was over Conant turned to me and said, 'Well, do you *want* to go to lunch?'"[43]

In truth, Conant and Skinner had never been particularly friendly. Their uneasy relationship dated back to the former's stint as the president of Harvard when the latter was a graduate student. "Somehow I disliked Conant," Skinner recollected of his time at the university under Conant's leadership.[44] "His repugnance," Skinner's biographer Daniel Bjork clarifies, "began in the mid-1930s when the just-appointed Harvard president refused to support Skinner's old mentor, William J. Crozier, in his plans to develop a general physiological department." Bjork also posits that Skinner didn't approve of Conant's proximity to political power: "His Manhattan Project position had put him in contact with Roosevelt and Truman. Eisenhower had appointed him ambassador. Skinner was sensitive about academics who were close to the centers of governmental power."[45] Yet Skinner seemed to crave some of that legitimacy too and was frustrated that political leaders refused to take his ideas about behavioral engineering seriously.

Even so, it's not clear why Skinner believed that Conant would find teaching machines—let alone *his* teaching machines—to be an appealing avenue for educational reform. Despite the popular narrative that Sputnik gave education reform a scientific and technological emphasis, Conant's report was not at all interested in a technical sort of shift to classroom practices. His prescriptions for change were entirely bureaucratic. As Carnegie's John Gardner observed in the introduction to *The American High School Today*, it

> may appear educationally conservative in not commenting on promising experiments in areas such as educational television, or on new approaches to the teaching of mathematics, physics, and languages. But Mr. Conant's standing as a forward-looking educator needs no defense. Over the years he has shown a lively interest in school and college experimentation. This report makes only passing reference to the innovations mentioned simply because they will have to pass the test of time. In the present report Mr. Conant concentrates on those improvements in curriculum and school organization which can now be adopted with confidence by any school system.[46]

As Skinner later wrote in his autobiography, Conant was asking questions like "How many class periods were needed for flexible schedules? What ratios of pupils to teachers and counselors were acceptable? What subjects should be taught, and when? How much homework should be required? But he was not asking what I thought was the important question: How can teachers teach better?"[47] Conant did not appear to be interested in the new learning science—or in the associated technologies—Skinner was championing.

Nevertheless, despite Conant's rather brutal criticism of the prose in Skinner's manuscript, Skinner left their meeting feeling encouraged—encouraged enough, that is, to write to Rheem immediately about a suggestion that Conant had made: test the teaching machines on poor Black children.

Conant "thinks that we are trying to do too many things," Skinner wrote to Rheem's director of planning. The feasibility of the teaching machine could really just be demonstrated with one high-profile successful project showcasing its transformative capabilities. To that end,

> [Conant] suggested that we go into the schools of Harlem or a Negro district in Chicago and teach beginning reading. At the moment, these students simply do not learn to read and the NAACP, a very powerful pressure group, is raising the devil with everybody concerned. They would certainly back a large-scale project, Conant believes. His proposal is that we select about a thousand elementary grade students, give them all intelligence tests, and then see how far we can go in a matter of two or three years in teaching them to read. At the moment, these students hang around school, sleep or raise hell, and never do learn basic reading skills. As Conant put it, if you could carry that group of students through sixth grade reading you would have produced a miracle, and the world be yours with respect to the further development of teaching machines. He considers this entirely a matter of strategy.

Even though research was ongoing in a number of settings—among the students using the machines at Harvard, for example, as well as in several secondary schools around the country—Skinner said none of that would "make the kind of noise a single experiment of this sort would make. Conant kept referring to the parallel with the atomic bomb. People in the field in the early days kept pointing out the uses of nuclear energy in industry, transport, and so on, but it was Conant who insisted they bear down on a single objective and devote all their powers to achieving it."[48] With the students of Harlem as his test subjects, Skinner could rescue them from a failing system and as an ed-tech savior, he could then convince the rest of the world of the effectiveness of his teaching machines.

Skinner wrote again to Conant in the new year, indicating that he had become "more and more interested" in the possibilities of a Harlem experiment and that he was going to try to find financial support from the Office of Education to fund it.[49] But to his dismay, it appeared that he had completely misunderstood Conant's support—or perhaps, as Skinner's biographer puts it, Conant "seemed to not only renege on the project but never to have been fully convinced of its viability."[50] "I am afraid that I am not in a position to be of help to you along the lines you suggest," Conant responded curtly. "I am not in touch with the New York people in such a way as to enable me to approach anyone with the project you have in mind."[51]

"You may recall that my suggestion was something very different and involved two steps," Conant continued. "The first was for you to put forward in brief form the evidence that you now have that something could be accomplished by the use of teaching machines in connection with reading in the first three grades." Conant said that he needed first to see evidence that the teaching machines "worked" at all before helping Skinner find a school with a Black student body where he could run more experiments. "I was then going to submit this material without comments to a reading specialist in another city where there is a large Negro population and get his reaction," Conant clarified. "If his reaction was favorable, I was then going to suggest something on the scale you mention in your letter of January 14. Without the first step, I do not feel I am in a position to be of much assistance to you." Even with evidence that programmed instruction was effective, Conant said he doubted many districts would be interested. "I am wondering if you do not underestimate the resistance to new and novel procedures on the part of those who have the responsibility for the education of young children," he scoffed. "After all, if a new idea involving reading should turn out to be highly unsuccessful, the official who had authorized it would have the permanent damage to his 1000 pupils very much on his mind!"[52]

Conant clearly feared that something would go wrong with an experiment in Harlem. Students, already disadvantaged by the current school system, could be further damaged. More importantly perhaps, Conant's *reputation* could be damaged. But Skinner craved the attention that he imagined a Harlem "miracle" would give him and the legitimacy that Conant's endorsement of the experiment could provide. He wrote back to Conant, adamant that the idea was worthwhile. "We talked about so many things that I will not apologize for misunderstanding your suggestion." There was always the risk, conceivably, of damaging human subjects while conducting experimental research, Skinner admitted, but scientists like himself strove "to evaluate what will actually be done to maximize the probability that the results will be at least harmless, if not favorable. It was my understanding that this was what you were proposing and that you had selected a school system which is now failing miserably so that any change would be almost necessarily in the direction of an improvement."[53] How possibly, Skinner contended, could teaching machines be worse?

Skinner refused to apologize for misconstruing Conant's words. So, no surprise, his high regard for his own research expertise had no effect on Conant's position. Conant would not change his mind. He would not help.

Skinner wrote to Rheem with an update on the situation:

> As I indicated to you on the phone, I received a rather surprising letter from Dr. Conant raising the question of whether there is a danger that young students might be "damaged" by machine instruction, and suggesting that preliminary work be carried out first before entering upon the large scale operation he had earlier suggested. Evidently when he began to think of what was involved in presenting such a project to a school system, he began to anticipate the objections which might be urged by a public official anxious not to make any mistake. Although I don't think there is a ghost of a chance that any damage could be done, it will probably be politically advisable to begin on a smaller scale with local groups where we can convince the authorities simply by telling them what we plan to do, that there is no danger. We should then acquire a substantial body of evidence which could be used in undertaking a large scale experiment.[54]

The endorsement of Conant, arguably the best-known educator in the country, was not at hand—an endorsement that Rheem had no doubt hoped would boost sales and that Skinner had hoped would legitimize the teaching machine movement in a way that no "substantial body of evidence" ever could.

Just as Conant was hesitant to attach his name to the teaching machine movement, Skinner had also become increasingly concerned about how his own reputation might be adversely affected by an association with Rheem.

Skinner was worried that Rheem was ignoring not only his suggestions, but also the inquiries of other educators and psychologists who'd written expressing interest in the company's teaching machines. Skinner shared an excerpt of what he said was a representative piece of correspondence with Rheem's director of planning. "I want to give you some details of my dealings with the Rheem corporation as I am very distressed with the way in which they operate," another researcher had told Skinner. "I have phoned them and written them, but I just cannot seem to get to first base with that outfit. They just don't seem to want to sell us any machines."[55] Skinner feared his own professional stature was now tied to Rheem's performance, and in early February 1960, Skinner wrote to Rheem alarmed with how the company was using his name and likeness to talk about its teaching machine products.[56]

If anything, Rheem was not damaging Skinner's reputation as much as it was squandering any commercial advantage the association might have given the company. Rather than, as Skinner feared, using the psychologist's name and likeness to sell a line of shoddy teaching machines, Rheem still seemed uncertain as to whether or not it wanted to manufacture any machines at all.

Skinner sent I. G. Davis a long letter in April, complaining about what he considered Rheem's failure to live up to the terms of their initial contract. "I entered into an agreement with Rheem rather than with anyone of several other companies because of the assurance that they intended to proceed on a broad front in developing teaching machines,"[57] he said, itemizing all the problems with the various machines that Rheem had under development—the "pre-verbal" machine, the "recall-by" machine, the "recall write-in" machine, and the "write-in" model that Rheem had supplied to several schools for experimentation. "I have been repeatedly assured that Rheem was soon to develop machines on the scale suggested by our agreement. . . . The simple fact is that Rheem has not yet built so much as an acceptable engineering model of anyone [*sic*] of the machines specified in our agreement of July 1, 1959." Skinner insisted that he was not requesting a change in the terms of his agreement with the company, but "simply that Rheem live up to it. I'm aware that Rheem must be responsive to changing market conditions but nothing has happened since the conclusion of our agreement to make the prospect of teaching machines any less bright. On the contrary, recent results have exceeded the most optimistic predictions."

If Skinner had hoped that his letter would be well received or would prompt a positive change in Rheem's behavior, he was mistaken. Davis wrote back on May 6, accusing Skinner of distorting the terms of their agreement. Rheem had not agreed to anything that would force it to develop teaching machines in a certain way or on a certain timeline. "Specifically, we feel we must restate that contrary to the implication of your letter, Rheem has not committed itself in its agreement with you to do anything in the teaching machine field after 1959 except for the payment to you for consulting services from time to time rendered." Davis accused Skinner of demanding more than his contract stipulated—the rights, for example, to approve any design and marketing of teaching machines Rheem built. "We feel it necessary to point out that under our agreement, Rheem is free to sell any teaching machines which it may feel should be marketed whether or not you agree that its design makes it psychologically appropriate," Davis wrote. "In fact, if you do not feel that the machine utilizes your 'procedures, techniques, research findings and scholarship,' while we would be unable to associate your name with the marketing of this machine we would nevertheless be able to market it and would not in such case be accountable to you for royalties under the royalty schedule with respect to such machines."[58]

Skinner forwarded the letter to his attorney in a state of panic. "If they are right, my hands are completely tied. They can simply manufacture machines I don't approve of, even though quite similar to my original designs, and can avoid all royalty payments." He added sarcastically: "Nice people!"[59] Rivkin wrote back to Skinner the next day, agreeing that he also found the Rheem letter "disturbing." He concurred that the company seemed to be seizing on the psychologist's frustrations in order to get out of having to pay royalties on the devices. Rivkin urged Skinner to clarify the matter with Rheem: "I think you would be well advised to make it clear that you do not think that the letter correctly states either the letter or the spirit of your agreement." Skinner also needed to remind Rheem that "the machines they are now developing or producing are indeed based upon and utilize your procedures, techniques, research findings and scholarship . . . but that their design and engineering is defective."[60] Rivkin helped Skinner craft a response to Rheem. "I am glad to confine myself to psychological principles," Skinner wrote. "I had never supposed that my comments on the design aspects of the machines displayed last September and my suggestions respecting particular engineering features of presently-projected machines were necessarily 'binding' on Rheem."[61] The immediate crisis seemed to be smoothed over.

In July, Skinner wrote to Rivkin once again, complaining that Rheem seemed to be tooling up for production without having let him look at the engineering model. "However, if that's the way they want to work," he said, "I am content to let them since I am trying to take things easy."[62] But that lackadaisical attitude seemed to belie how he really felt. Ten days later, he wrote to Rivkin again, grumbling that he still hadn't received his July royalty check. He suggested Rivkin swing by the offices of Rheem's attorney, Walter Lewis, in New York to try to resolve things, but to "please keep in mind that I should not be at all sorry to terminate my working relationship with the Rheem Company."[63]

Skinner wrote again to Rivkin a little over a week later; he'd still not received a check from Rheem. And he had more issues that Rivkin could bring up with Lewis, should he choose to stop into his office: he'd sent over six hundred names of potential customers to Rheem, but none of those people seemed to have heard from the company; nor had Rheem asked for Skinner's feedback on its latest device. "I have no way of knowing that the final production model will not contain some egregious bloomer [*sic*] which would seriously limit its usefulness. It seems to me ridiculous that a company should not make use of its consulting services at every stage during the development of a piece of equipment. These are just samples. The main complaint is that Rheem has not done what it promised to do a year ago. It is now seriously undertaking to make only one simplified version of the three machines spelled out in our agreement."[64]

As Skinner hoped, Rivkin met with Rheem's general attorney in August. Writing to Skinner with details of the meeting, Rivkin said that Walter Lewis admitted most of Skinner's grievances were justified. "He was at pains to say that the Califone transaction had created great administrative confusion. He also said that the technicians from Hughes Aircraft that Rheem had engaged did not have the qualifications which were hoped for. As a result of the managerial chaos that obtained, he said, Rheem is now about nine months behind its projected schedule."[65] Rivkin assured Skinner: Lewis was emphatic that Rheem remained interested in the teaching machine business and was "still enthusiastic about its affiliation with Skinner."

Rheem executives did have concerns, however. They felt that the development of programming materials lagged far behind that of the machines. They were also unhappy about the lack of standardization among the various manufacturers with respect to how they'd hold or handle programs. And many at Rheem anticipated that electronic teaching machines would soon replace mechanical ones. As such, there were some at the company who thought it might not be an appropriate time for Rheem to commit itself to full-scale production of teaching machines. Rivkin told Skinner that he'd ended the meeting by showing Lewis an article from the front page of the *Wall Street Journal*. "I said that I thought it was tragic that Rheem and Skinner are mentioned briefly and near the end of the article [in] which a bunch of Skinner students and disciples and the companies they now work for are mentioned prominently and at the head of the article. Rheem, it seemed to me, had dissipated the enormous competitive jump it had in the field."[66]

Skinner's emotional seesaw continued. Sometimes he was encouraged and sometimes he was despondent about the prospects for his work with Rheem. In late September, Skinner wrote to Rivkin informing him that "the top people in Rheem Califone have been here all morning, and I really think they mean business at last, though God knows why I believe anything they say."[67] The Rheem executives had informed Skinner that they'd set a November 2 deadline for a trial run of 150 machines, which they planned to use as a pilot program in Los Angeles schools.

A manager in Rheem's marketing department asked Skinner to visit the LA experiment to see "whether it was being done in a scientific fashion." But Skinner said he didn't think he could make the trip. Ironically, perhaps, he echoed James Conant's concerns about the scheme to run an experiment on the students of Harlem; it was less a question of logistics than it was a matter of Skinner's fear that if he visited the schools and found that the experiment wasn't well designed or that the machines were not functioning properly, it would be too late to do anything and "I would be on the spot professionally if the schools asked my opinion."[68]

Skinner was trapped. He needed Rheem to succeed with its plans to build his teaching machines. But he was no longer confident he should attach his professional reputation to the company's manufacturing efforts. As 1960 drew to a close, the patent for the new Didak 501 had not been completed. Indeed, the *design* of the machine had not been finalized. Rheem still did not have a product that was commercially available. And that meant that the famed psychologist—the man whose name was most closely associated with the teaching machine—didn't either.

6

PROGRAMMED INSTRUCTION: IN THEORY AND PRACTICE

Harvard had given B. F. Skinner a large room in the Batchelder House on campus in 1954 where he and his team of "bright young behaviorists"—a group that initially included Lloyd Homme, Susan Meyer (Markle), Douglas Porter, Irving Saltzman, Matthew Israel, and Wells Hively[1]—had started work on designing their new teaching machines as well as "programs," the materials that would accompany them.[2] "We had, of course, never seen an instructional program," Skinner admitted. "How much of a subject should it cover? How much in a single session? How much in each 'frame' (as we began to call each presentation)? If frames were to reappear for review in later parts of a program, how should they be distributed? How much could we assume students already knew, and where were we to find students who were at the right point to a test a program?"[3] These questions were central to the development of this new instructional technology—indeed, to the very *idea* that there was such a thing as a scientific practice of instructional design.[4]

By the end of the year, Skinner and his teaching machine group had developed short programs in "kinematics, trigonometry, coordinate systems, basic French words and material to teach French dictation, phonetic notion, vocabulary and rudimentary grammar, as well as single demonstration disks in geography, anatomy, and poetry."[5] But Skinner recognized that, despite having four "programmers" and two graduate students on his staff, he could not afford to hire people with enough expertise in these and other fields to develop longer or more elaborate programs—the kind, say, that could be used to teach an entire college-level course. So, Skinner decided that *he* would serve as the "content expert." He would use teaching machines for his own class, Natural Sciences 114, a general education course in which he taught behaviorism. He and his team would write a teaching machine program based on his 1953 book *Science and Human Behavior*. Teaching machines for the course were installed in a "self-instruction room" in the basement of Sever Hall.

The results of this experiment with teaching machines were encouraging, Skinner and his team declared, and students "reported rather favorable impressions of the machine work," with 62 percent of those students who used the devices indicating that programmed instruction made the coursework easier to understand.[6] Enrollment for Natural Sciences 114 jumped 70 percent in one year, an increase that Skinner chalked up to the popularity of the teaching machines.[7]

Particularly significant to this early work in Skinner's lab was Susan Meyer (Markle),[8] who'd been hired to write the arithmetic program for the IBM machine. Meyer Markle's role was often minimized, even by Skinner, who remarked in his autobiography that "the secretarial role . . . fell to her according the standard of the time as the only woman in the group."[9] In a letter Meyer Markle wrote to Skinner decades later, she talked about how these sorts of assumptions had served to diminish her stature in the field she'd helped found, and she noted that, despite being the first to publish a dissertation on programmed instruction, another man in the field, James Evans, "laid claim to being the first, until, as sexism went down the drain, he asked the date of mine."[10] She was one of the only members of Skinner's group at Harvard who was regularly outside the lab and "in the field," working directly with schools to write programs, to implement their teaching machine projects, and to assess their efficacy.

While Skinner might have introduced the idea of programmed instruction to the public, Meyer Markle helped establish many of its conventions—prompting, fading, and so on.[11] She ran numerous workshops with teachers in order to create and refine teaching machine materials. She also published extensively on the ideas of programming and instructional design, including the books *A Programed Primer on Programing* (1961) and *Good Frames and Bad: A Grammar of Frame Writing* (1964).[12] (Meyer Markle insisted that "programmed instruction" have only one "m.")

Programmed instruction was poised to transform education from an art to a science, Meyer Markle asserted. It was not an entirely new argument; advocates for standardized testing had long contended that the precision of new academic measurements would do the same. But programmed instruction was not merely assessment that occurred at the *end* of the lesson to gauge student learning; it was the entire lesson itself. The program "is the controlled environment in which learning is to take place," Meyer Markle wrote. "Every step that the student is to go through is there, on paper and on tape. The teacher-programer knows exactly what is happening to the student." That is, even when the programmer was not present, so well-engineered—ideally, of course—was the program, so controlled was the learning environment, that the student's responses could be predicted and his errors understood. The program, Meyer Markle argued, "gives information to the student and gets from his responses at each step indicating that he has understood this information. If he does not answer questions correctly, the teacher knows that something has gone wrong in the communication process. On the basis of what went wrong, a change in the controlled environment can be made. The new conditions are then tested for their effect on students." When properly constructed and implemented then, programmed instruction would require perpetual revisions to the programming materials—certainly something that would create *more* work, not less (although work for the engineer, not necessarily for the educator). "The result is an applied science of textbook writing, in which the texts are tested sentence by sentence by the students for whom they are designed. The applied scientists, the programers, vary and revise and reshape the program until it produces the designed result—learning."[13]

Programmed instruction was *individualized* instruction. Meyer Markle likened it to the work of a tutor, "a master of intellectual teasing" who adjusts the lesson to her student's needs but also challenges the student to keep moving forward.[14] If the tutorial relationship was the ideal—something that many educators, often invoking the ancient Greeks, seemed to believe—then programmed instruction sought to become the technological version of this: "Each student was now to have his own private tutor, encased in a small box," Meyer Markle wrote.[15]

Despite Meyer Markle's role in developing some of the core concepts of "programing" and her presence from the outset in Skinner's teaching machine group, she was rarely profiled by the press or featured in histories of teaching machines, except when mentioned as a graduate student tasked with working with his IBM machine. Perhaps that's because her work was with the "software" and with teachers and students, not with "hardware" and not with industry. (Although she did develop an arithmetic program for IBM, recall that Skinner retained the rights to her work.) Or perhaps that's because the story of education technology tends to prioritize men and their machines.

More likely to receive credit for innovations in programmed instruction was Norman Crowder. In 1958, he published a "scrambled textbook" titled *The Arithmetic of Computers*.[16] The book was the first in a series of self-instruction manuals—"TutorTexts"—published by Doubleday. (Other titles in the series taught algebra, trigonometry, electronics, and the game of bridge.) "Now you and your family can have a private tutor in your own home to help you learn complicated math subjects with push button ease," an advertisement in *Popular Science* pronounced, trumpeting, "TutorText: the revolutionary Age of Automation innovation in self-help books."[17] The TutorText was "a complete programmed teaching machine in book form," the ad explained, an early indication that publishers would try to incorporate programmed instruction into their catalogs, even without actual teaching machines to accompany them.

The TutorText was "a book written by a new technique developed through recent advances in automatic teaching methods," the preface of Crowder's 1958 book began. "The presentation of material in this book approximates, as nearly as possible, a conversation between a teacher and his pupil," it read, again invoking that tutorial ideal. The book provided lessons in "small units," followed by multiple choice questions "which the reader must answer in order to proceed further in the book. A wrong answer leads to more discussion of the same point of information; a correct answer leads to the next unit of information and the next question."[18] The book contained more traditional, end-of-chapter quiz questions too, as it was meant to serve as an entire course of study. "The reader's rate of progress through the course is determined only by his facility for choosing right answers instead of wrong ones. It is not recommended, however, that the book be read in one sitting, or even in two or three. A number of shorter learning periods produce better results than one long session," the preface advised.[19]

A TutorText was not meant to be read cover to cover, but rather the reader would move through its pages based on whether or not she could answer the questions at the bottom of each one correctly—sent to page 45, for example, if she got the answer right; sent to page 73 if she got the answer wrong and needed a concept explained again in a different way; or sent to page 78 if she got the answer wrong because she'd made a simple error in calculation. In this way, Crowder argued that this method of offering alternative routes through the material rendered the TutorTexts "adaptive," taking a student through a particular, personalized path based on their answers—a "Choose Your Own Adventure" of sorts, the name of a beloved series of fictional books that were published beginning in the late 1960s and modeled, if not directly then certainly indirectly, on these popular "scrambled textbooks."[20]

Crowder called his version of programmed instruction "intrinsic"—although it was often described as "branching" to differentiate it from Skinner's more "linear" model in which all students proceeded through the same questions. "Linear and intrinsic programing have nothing in common historically," Crowder insisted, "having arisen in different circumstances. They have nothing in common theoretically, but rather rely for their expected effectiveness on different rationales and make different, and in fact, diametrically opposed assumptions about the nature of the learning process."[21] In the linear model of programming, Crowder argued, the learning *theory* came first—that is, it emerged from Skinner's behavioral science. The intrinsic model, on the other hand, was based on a new *technique*, unencumbered by any theoretical presuppositions.

"The student is given the material to be learned in small logical units (usually a paragraph, or less, in length) and is tested on each unit immediately," Crowder explained.

> The test result is used automatically to conduct the material that the student sees next. If the student passes the test question, he is automatically given the next unit of information and the next question. If he fails the test question, the preceding unit of information is reviewed, the nature of his error

is explained to him, and he is retested. The test questions are multiple-choice questions, and there is a separate set of correctional materials for each wrong answer that is included in the multiple-choice alternative. The technique of using a student's choice of an answer to a multiple-choice question to determine the next material to which he will be exposed has been called "intrinsic programming."[22]

Unlike Skinner, who believed in minimizing errors and in forcing students to compose rather than select their answers, Crowder believed that errors in multiple-choice scenarios could be useful in diagnosing *why* a student made a mistake. Crowder also pushed back against Skinner's insistence that each frame should display the smallest possible "step." This often meant the program was too easy, and Crowder claimed it regularly led to programs that were simply boring.

"Automatic tutoring by intrinsic programming is an individually used, instructorless model of teaching which represents an automation of the classical process of individual tutoring," Crowder wrote.[23] While Skinner and Pressey were quick to insist that their teaching machines would not replace teachers, Crowder clearly felt less obligated to do so. As his materials were marketed mainly to the adult learner at home and to major corporations interested in training employees, there was less concern, in Crowder's rhetoric at least, of offending teachers by suggesting that programmed instruction would replace them.

With his crew cut, dark spectacles, and bow tie, Crowder was stylistically, not just scientifically, a very different figure from Skinner, whose red hair had turned grey by the late 1950s but retained its curly flop. "He writes in a chatty, colloquial vein," one journalist described Crowder, contrasting him with the older Harvard psychologist whose explanations often relied on the jargon of behavioral science.[24] Skinner was fairly disdainful of the younger man and openly dismissed his work as, if nothing else, insufficiently behaviorist—a charge that was not entirely fair but was quite representative of Skinner's treatment of others in the field. Crowder did frame intrinsic programming as "modifying behavior (exposing new and different material to the student) until the desired result is obtained" and he held a PhD in psychology.[25] But he was not an *academic* psychologist—certainly not one with as prestigious an affiliation as Harvard—and he did not move in academic circles; rather, Crowder had been a military psychologist before being hired as an industry one. And the power and influence of his employer, Western Design, a subdivision of U.S. Industries, was not insignificant. Even if his job did diminish him in Skinner's eyes, Crowder was able to muster quite an impressive public relations presence. Crowder's version of programmed instruction was often framed—by other psychologists as well as by the press—as being *the* alternative to Skinner's.[26]

For her part, Susan Meyer Markle was much less dogmatic than Skinner, and she welcomed some of Crowder's ideas, admitting in a 1962 lecture that "I think we are going to arrive at the conclusion in the near future that the simple linear sequence is inadequate."[27] In a subsequent talk, this one delivered at the 1964 convention of the National Society for Programmed Instruction, Meyer Markle recognized, "It has become fashionable of late—as late as yesterday, for instance, to attack the so-called formulae of linear programming as archaic birdseed, unfit for human consumption. We have heard complaints from the granddaddy of us all about this modern generation—with its emphasis on the shiny chrome of reinforcement and the push-button ease of errorless learning. The blame, of course, is laid squarely at the feet of the felt proximity of the pigeon lab to the teaching lab on the Harvard campus."[28] She argued that, based on the notes she'd been tasked with taking as part of the Harvard teaching machine project, she could see that some members of Skinner's team were already thinking about "branching" well before Crowder gained recognition for the idea of presenting students alternate paths along which to proceed through the programmed materials based on their right or wrong answers.

The debates about the right direction of teaching machine research and development had emerged soon after Skinner made his first presentation on the topic in 1954. By the late 1950s, several academic conferences on the topic of teaching machines had been held—an indication that even without the commercial production of a Skinner device, there were plenty of other scholars who were actively trying to shape the field. (There were plenty too who were readily trying to resist it. At one demonstration of a teaching machine to the University of Michigan Department of Educational Psychology in 1958, things turned "just short of violent."[29])

One of the first academic conferences on teaching machines was held in December 1958, when the University of Pennsylvania and the Air Force Office of Scientific Research convened a two-day meeting on the "Art and Science of Automatic Teaching of Verbal and Symbolic Skills." Among those who presented papers were B. F. Skinner, Susan Meyer Markle, Lloyd Homme, Norman Crowder, and Sidney

Pressey. While most were eager to showcase their research and discuss the benefits of teaching machines, Pressey was unimpressed. Indeed, he was "startled," he later wrote, "by the learning theorists' ignorance of the great amount and variety of research regarding learning in school and assurance in applying there [sic] concepts derived primarily from rat maze-running or paired associate memorizing."[30]

When A. A. Lumsdaine and Robert Glaser compiled their edited collection *Teaching Machines and Programmed Learning* a few years later, they seemed to do so partly in response to Pressey's concerns about the ahistoricism permeating the field. (Both men attended the Penn conference.) Access to research—by the public and by scholars—was limited, and this, they feared, was likely shaping how teaching machines were being developed and assessed. Lumsdaine and Glaser wrote in the introduction to their book:

> Educators and psychologists, in increasing numbers, have been seeking information about activities in this field. This effort has encountered considerable difficulty because of the relative inaccessibility of much of the material. Published articles in this field have appeared in a widely scattered periodical literature. A number of the more recent contributions have not been published. Dissemination of information has depended largely on word of mouth and personal correspondence. This has been quite an inefficient process, involving much overlapping of effort, and has also imposed a heavy burden upon the authors of papers. Available supplies of reprints of copies of unpublished papers have frequently been exhausted.[31]

The papers from the Penn conference had been published in a 1959 book edited by Eugene Galanter—the only book available on teaching machines, Lumsdaine and Glaser pointed out, when they took theirs to press.[32]

As such, they hoped to publish something that would offer a broader introduction to the history of teaching machines, not just a survey of the most recent research and theory. The volume included the three articles that Sidney Pressey had written for *School and Society* in the 1920s, along with a reprint of B. F. Skinner's 1958 article in *Science*. It contained a dozen articles by Skinner and the members of his teaching machine group at Harvard—Douglas Porter, Lloyd Homme, James G. Holland, Wells Hively, and Susan Meyer Markle—as well as research from professors from Hamilton College, Earlham College, the University of Utah, and UCLA. The book also provided a number of contributions from those working on teaching machines for the military and for industry, including an article by Norman Crowder and two articles by Leslie Briggs, a student of Sidney Pressey, who had gone on to work for the US Air Force and the American Institutes for Research. At almost eight hundred pages, Lumsdaine and Glaser's "source book," published by the National Education Association (NEA), became one of the most heavily referenced books on the topic, helping to establish the key names and narratives of the burgeoning teaching machine movement.

Importantly, the book served to delineate what "counted" as a teaching machine, quite a task considering Skinner's loud objections to those who he believed did programmed instruction wrong—a long list that included Sidney Pressey and Norman Crowder. Lumsdaine and Glaser offered their own definition, arguing that teaching machines had three properties:

> First, continuous active student response is required, providing explicit practice and testing of each step of what is to be

learned. Second, a basis for informing the student with minimal delay whether each response he makes is correct, leading him directly or indirectly to correction of his errors. Third, the student proceeds on an individual basis at his own rate—faster students romping through an instructional sequence very rapidly, slower students being tutored as slowly as necessary, with indefinite patience to meet their special needs.[33]

Of course, Glaser, as a teaching machine businessman himself, had a horse in this definitional race, and the source book had to navigate the complexities of professional reputations (and egos), commercial interests, and the NEA's own position—as the largest teachers' union in the United States, that is—in creating an early outline for the field.

"The principles used in most of the programs constructed by Skinner, Gilbert, Homme, Glaser, Power, and others appear very reasonable," Lumsdaine and Glaser offered, "but have not yet been subjected to clear-cut experimental tests."[34] In 1960, much was still "indeterminate." And yet much was written—particularly by industry marketers and by journalists—as though all the questions about design, efficacy, and pedagogy had been settled.

This was the criticism that Sidney Pressey had lodged during his presentation at that 1958 conference: "It is not enough that in the experimental situation the proposed new methods work well. They must do so in the average situation where they are to be used and with average people there; and they must be sufficiently better than the methods and materials these same people have been using, that a changeover is both warranted and feasible."[35] Pressey worried that, by bypassing thorough classroom experimentation and relying instead on industry prototyping and marketing, teaching machines might head in entirely the wrong direction.

He wrote to Ben Wood that whatever research there was on teaching machines and programming was poised to be "irresponsibly exploited."[36]

His cautionary note was not heeded, and as Pressey later wrote, "I was shocked at what followed: the most extraordinary commercialization of a new idea in American educational history—hundreds of teaching machines were put on the market, some sold door-to-door with extravagant claims, others costing thousands of dollars, hundreds of 'programs' published with as many as 16,000 frames, all involving many millions of dollars of investment. Then millions of research dollars went into, first, the confident elaboration of these ideas and only slowly into any questioning of them."[37]

7

IMAGINING THE MECHANIZATION OF TEACHERS' WORK

Despite Sidney Pressey's hope that those building teaching machines would rein in their claims—claims that, to be fair, he himself had made when he heralded the "coming 'industrial revolution' in education"—advocates for teaching machines could not contain themselves, making grandiose promises about innovation and efficacy, whether or not their assertions were backed by research.[1]

Predictions that the future of education would be increasingly technological were hardly new. Thomas Edison regularly made them. For example, in 1913 he forecast that "books will soon be obsolete in the public schools," to be replaced by the motion picture.[2] But in postwar America, the technological imaginary truly thrived, as visions of robot teachers and mechanical brains and automated schools spread through popular culture.

"We are in rapid transition today to a new world which threatens to be dominated by technological advance,"

Simon Ramo wrote in 1957, in an essay in *Engineering and Science* published the same month as the launch of Sputnik.[3] Ramo was, at the time, the vice president of Thompson Ramo Wooldridge Inc., an aeronautics manufacturer in Los Angeles; he would later go on to design the first intercontinental ballistic missile. "In that new world," Ramo predicted, "man will have learned so much about nature's store of energy and its release that he will have the ability to virtually destroy civilization." Furthermore, "production, communications, and transportation will all be 'automatic'—these operations of man's material world will have become so vast and complex that they will have to proceed with a minimum participation by man, his muscles, brains, and senses." And finally, "man will conquer space." Automation. Nuclear weaponry. The Space Age. All this scientific advancement, Ramo argued, was a signal of "the coming crisis in education."[4]

> Already, the increasingly technical world uses more scientists and engineers, yet the very industrial development that is part of the growing technical society takes the engineers and scientists away from the university and high-school facilities, and the fast world in which we live makes the long study of science seem unattractive to the youngsters. The technical society is complex, rapid, and increasingly dangerous. We can blow up the whole world, yet such a premium is put on the use of our human and physical resources for everything but education that it seems that the new technical society is going to be accompanied by a weakened ability to keep pace education-wise.[5]

To prepare students for a technical society, education must necessarily become more technical, otherwise there would be no way for American society to keep up.

"Now, if the world were in transition to something different on a very, very slow scale, we could argue these factors would take care of themselves," Ramo continued. "Supply and demand would then presumably set to work to make the teaching profession pay off better," attracting more people and arguably better people to the job.[6] But the world was changing faster than it had ever changed before, Ramo argued, and that required a dramatic shift in schooling, one that could not wait.

Ramo proposed "a new technique of education," insisting that the wildly speculative scenario he described would better equip schools for the future than would, say, the gradualism of "greater use of television as a teaching aid," something that several districts were already exploring.[7] Instruction "through the eye" was, after all, Edison's prediction—a fantasy that was already decades old.[8]

"First of all," Ramo imagined, "we will get the student registered. I won't burden you with the details here; when the registration is complete and the course of study suitable for that individual has been determined, the student receives a specially stamped small plate about the size of a 'charga-plate,' which identifies both him and his program. (If this proves too burdensome for the student, who will be required to have the plate with him most of the time, then we may spend a little more money on the installation and go directly to the fingerprint system.)"[9]

This "charga-plate" would enable the individualization of education thanks to the vast amount of data it would contain about each student. Ramo explained: "When this plate is introduced at any time into an appropriate large data and analysis machine near the principal's office, and if the right levers are pulled by its operator, the entire record and progress of this student will immediately be made available. As a matter of fact, after completing his registration, the student introduces his plate into one machine on the way out, which quickly prints some tailored information so that he knows where he should go at various times of the day and anything else that is expected of him."[10]

The typical school day, Ramo imagined, would still consist of a number of classes—some with other students and some alone with a teaching machine. "Sometimes a human operator is present with the machine, and sometimes not."[11]

Ramo envisioned the student would spend a few hours a week studying a topic like trigonometry "in automated classrooms. In the case of trigonometry, only a small part of his time need be spent with a human teacher. Some of his classroom exercises will involve presentation of basic concepts in trigonometry in the company of other students in short lectures, delivered by a special motion picture, which uses some human actors who enunciate or narrate the principles to the accompaniment of various and sundry fixed and animated geometrical diagrams." These, Ramo said, would be "pushbutton classes"—the classroom furniture all designed to function as part of a giant teaching machine system. "Each chair includes a special set of push buttons and, of course, that constant slot into which the student places his identification plate. The plate automatically records his presence at that class, and it connects his push buttons with the master records machine."[12]

"If the class is large," Ramo argued, "our student is much less likely to sleep or look out of the window than in a normal lecture by a human teacher."[13] The student would be peppered with quiz questions, and he would continually be prompted to press various keys to make sure he was paying attention and understood the lesson. The push-button system was interactive, Ramo insisted, implying that pushing and clicking on buttons made for a more engaging form of learning.

Although these large classrooms that relied on film-based instruction might seem like "mass" education, the student would always receive individualized instruction as well. The student, Ramo said, "is in constant touch with the 'teacher'"—or at least with a teaching machine. The data collected by the machine would be "used by the electronic master scheduling device to prepare for the special handling of that student." During that portion of the instruction, the student would be

> seated in front of a special machine, again with a special animated film and a keyboard, but he is now alone and he knows that this machine is much more interested in his individual requirements. It is already setup in consideration of his special needs. It is ready to go fast if he is fast, slow if he is slow. It will considerably repeat what he has missed before and will gloss over what he has proven he knows well. This machine continues the presentation of some principles and asks for answers to determine understandings. Based upon the student's immediate answer, it may repeat or go on to the next principle. With some hints and assistance by the lecturer in the movie, and with appropriate pauses (not accompanied by a commercial), the student is allowed a period for undisturbed contemplative thought before registering his answer.[14]

The machine would be adaptive, Ramo imagined, "prepared to take a single principle and go over it time after time if necessary, altering the presentation perhaps with additional detail, perhaps trying another and still another way of looking at it, hoping to succeed in obtaining from the student answers that will indicate that the principle is reasonably well understood before it goes on to the next one."[15] Thanks to the pervasive data collection and analysis, the machine would know precisely what the student needed before he sat down at it.

"A brilliant student could romp through trigonometry in a very small fraction of the course time," Ramo maintained. "A dull student would have to spend more time with the machines. The machines can be so set up that if a student failed to make progress at the required rate, he can automatically be dropped from the course. Of course, before that happens or before the brilliant student is allowed to complete the course, a special session with that student by a skilled teacher is indicated. But the teacher will be aided by having before him the complete records of what could be weeks of intensive machine operations." This, said Ramo, would enable the long-sought individualization of education—"a personal study of that student's understanding and his way of thinking about a subject."[16]

Ramo, like many promoters of teaching machines, insisted that the introduction of automation would alleviate the drudgery of certain kinds of teachers' work and would enable more time for more individualized attention. "It is for this reason, although we can use motion pictures and television to replace a lecturer and can, in theory at least, be more efficient in the use of one skilled teacher's time, enabling him to reach a larger audience," Ramo admitted, "we can only use such techniques for a limited fraction of the total school day. . . . The whole objective of everything that I will describe is to raise the teacher to a higher level in his contribution to the teaching process and to remove from his duties the kind of effort which does not use the teacher's skill to the fullest."[17]

Ramo believed that this "new technique of education" would create "a new profession known as 'teaching engineer,'" he speculated, "that kind of engineering which is concerned with the educational process and with the design of the machines, as well as the design of the material."[18] (Carnegie Mellon professor Herbert Simon would, a decade later, make a similar call for a "learning engineer" to help combat the "amateurism" of educators in designing efficient "learning environments."[19]) Ramo's "new technique" would also spawn a new industry, devoted to the manufacturing of educational machinery, the lessons they would display, and the data storage and analytics programs the push-button school would require.

All this would change the labor force (and labor practices) of the school. For despite all the insistence that these machines would serve to enhance the role of the teacher, much of the teaching, by Ramo's own admission, would be handled by an "operator," not an educator. Ramo foresaw that "the high school becomes partially transformed into a center run by administrators and clerks, with a minimum of the routine assigned to the teaching staff. The teaching staff is elevated to the role that uses the highest intelligence and skills. A smaller number of teachers makes possible the education of a larger number of pupils."[20]

"From the standpoint of the student," Ramo said, "I do not know that his life need be changed in any fundamental way. It may be, of course, that the evenings and weekends would cease to be times for doing homework. The equivalent of homework, as well as the basic presentation periods, would be done perhaps during the normal working day, five days a week, with the evenings and the weekends used for the broader cultural, social, and athletic events."[21]

In their children's book *Danny Dunn and the Homework Machine*, published the year following Ramo's essay, Jay Williams and Raymond Abrashkin seemed to disagree that nothing much would change for students. Indeed, they implied that automated education would actually make students work more. In their story, the titular character Danny and his friends use the computer of their neighbor, Professor Bullfinch, to automate the completion of their homework. When their teacher discovers what they've done, they're accused of cheating. Rather than praising them for using the machine as a labor-saving device, to eliminate the drudgery of their lessons, their teacher tasks them with even more homework.[22]

Simon Ramo's fantasy future inspired several other writers and artists, including Arthur Radebaugh who drew the cartoon "Push Button Education" based on the essay, appearing in May 1958 in his syndicated comic strip *Closer Than We Think* (see figure 7.1).[23] It's worth pointing out that both Ramo's 1957 essay and Radebaugh's comic came out *before* B. F. Skinner's article on teaching machines was published in *Science* (in October 1958). And *Closer Than We Think* boasted some nineteen million readers—vastly more than *Science*.[24] Arguably then, the popular notions of the future of education—teaching machines and push-button

IMAGINING THE MECHANIZATION OF TEACHERS' WORK 157

Figure 7.1
Arthur Radebaugh's cartoon "Push-Button Education" used with permission from the Tribune Content Agency.

schools—were shaped, initially at least, less by the Harvard psychologist and more by the colorful comic-strip imaginary.

Not surprisingly, A. A. Lumsdaine and Robert Glaser included Ramo's popular essay in their massive book on teaching machines, *Teaching Machines and Programmed Learning*. While the two had lamented that there was just one other book on teaching machines when they published theirs in 1960, the number of titles about the field grew dramatically in the years that followed, as scholars and journalists and cultural commentators sought to explain and promote the idea of automated education: *Teaching by Machine* (1961), *The First Book of Teaching Machines* (1961), *Programmed Learning and Computer-Based Instruction* (1961), *An Introduction to Programed Instruction* (1962), *Programed Instruction: Today and Tomorrow* (1962), *Programmed Learning and Teaching Machines: An Introduction* (1962), *Teaching Machines and*

Programed Instruction: An Introduction (1963), *Programs, Teachers, and Machines* (1964), *Programmed Learning: The Roanoke Experiment* (1965), *Teaching Machines and Programed Learning: Data and Directions* (1965), and *Programmed Teaching* (1965), for example.[25]

A similar pattern occurred in magazines, which began covering the teaching machines in the late 1950s, sometimes citing the research that appeared in scholarly journals and sometimes quoting the staff at the schools that were early adopters of the automated education, but more often writing with a breathless excitement unmoored from either research or implementation. In 1960, there were articles in *Time*, *Business Week*, *New York Times Magazine*, *Fortune*, and the *Saturday Evening Post*, for example. In 1961, there were articles in *Popular Mechanics*, *Parade*, the *Christian Science Monitor*, *Science Digest*, and the *Commonweal*. In 1962, the *Reporter*, *Popular Science*, *Good Housekeeping*, *Look*, and *Changing Times* (later *Kiplinger's Personal Finance*) published stories on teaching machines. The topic was covered, as this partial list suggests, in women's magazines, business magazines, technical publications, and general news outlets. The stories—the promise of educational transformation by machine—were almost inescapable. Skinner and his teaching machines were featured in the CBS television show *Conquest* in 1959 in an episode on learning, behavior, and "what makes us human."

Predictions about the coming machine revolution were everywhere—and not just in the speculative fiction. "It will probably take from four to five years for programmed learning to become solidly established in public education, but I have no doubt that it will," one teaching machine enthusiast told *Popular Mechanics* in 1961.[26] The following year, in an article in *Popular Science*, Norman Crowder predicted that by 1965, half of all students would be using teaching machines, "at least for a course or two."[27]

Many of the articles claimed that teaching machines would greatly enhance students' interest and students' pace of learning. According to one article,

> In an experiment sponsored by the U.S. Office of Education, sixth-grade pupils were taught spelling by machine for a period of six months. Although they spent only a third as much time on spelling as their classmates who were being taught in the ordinary way, they scored much higher on standard achievement tests. In Roanoke, Virginia, eighth-graders of average ability learned just about as much algebra in one semester from a programmed text as ninth-graders are ordinarily expected to learn in a year. At a boarding school near Philadelphia, mentally retarded teen-agers were provided with machines designed to give them practice in arithmetic; at the end of the school year, tests showed gains in proficiency two and a half times as great as gains made by students in a closely matched control group who had not used machines.

And it wasn't just K–12 students who were learning faster. The same article said that "IBM has been able to reduce from fifteen to eight hours the class time needed to cover the opening sections of a course the company gives on the use of its 7070 computer." The research was overwhelmingly positive, according to the press coverage. "Experimenters also report, as a rule, that students like programmed instruction and think it does them a lot of good."[28] *As a rule*—that is, there was no other way to think about the future of education than as one that would be programmed in this way.

That teaching machines worked better and faster than human teachers was certainly a story that appealed to the readers of business magazines, which seemed more than happy to repeat a story that derided the school system for its backwardness, its inefficiencies. This stance had found a friendly audience in the business community at least since the publication of Frederick Taylor's *The Principles of Scientific Management* in 1911. (This was the observation Raymond Callahan made in 1962 when he published his book *Education and the Cult of Efficiency* on the efforts in the early twentieth century to run schools like businesses.[29]) In a 1958 article, *Fortune* complained of "The Low Productivity of the 'Education Industry,'" blaming teachers and teachers' unions that the "output" of schools had not kept pace with investment.[30] Opening with the cliché that education was "big business," *Fortune* columnist Daniel Seligman sneered that teachers "oppose anyone who tries to apply business concepts to their work. The concept of productivity—i.e. output in relation to input—is especially abhorrent to educators, possibly because most productivity figures tend to make the education 'industry' look bad."[31] Teachers, Seligman contended, were so inefficient, they had no right to demand an increase in pay. Suggesting that in previous decades, schools were actually *more* productive—in part because of larger class sizes, he claimed that "thirty years ago students were educated more 'efficiently' than they are today, i.e. each student required fewer teaching man-hours—and fewer administrative, clerical, and custodial man-hours—than he does today. There is now one teacher for every twenty-six students, in 1928 there was one for every thirty students, and in 1900 there was one for every thirty-seven."[32] New technologies were going to change this, Seligman argued—whether teachers liked it or not. Mocking the teachers' unions' concerns, he likened their stance to "the locomotive firemen's union's early reaction to the diesel engine."[33]

To underscore how educational technologies were positioned to displace teachers, Seligman touted the adoption in 1956 of television-based education in Hagerstown, Maryland, "where 18,000 pupils, from the first through the twelfth grades, are receiving some instruction by television. The instruction is transmitted on a closed circuit from six 'studios' in Hagerstown at the rate, currently, of 120 sessions per week, to 450 classrooms equipped with conventional 21-inch black-and-white table models."[34] The cost savings, Seligman argued, indicated that "classroom TV is certain to pay for itself at the very least" by rendering teachers in the district superfluous.[35]

Two years later, *Fortune* devoted an article to teaching machines specifically—this one, penned by science journalist George A. W. Boehm. The article hailed programmed instruction as "the most radical innovation in education since John Dewey introduced his 'progressive' theories more than half a century ago." Indeed, programmed instruction seemed likely to be just as controversial as Dewey's ideas, Boehm speculated.

> The familiar routine of school—lectures, textbook study, recitation, regular quizzes, and even, to a certain extent, teachers—has been all but eliminated. Students work with printed "programs" designed to be so easy to follow that they can proceed almost without supervision and at their own pace. Programed teaching, if it lives up to its early promise, could in the next decade or two revolutionize education. It may also have an important impact on such U.S. educational problems as the shortage of teachers and the construction of schools. Conceivably it could upset the whole social structure of American youth.[36]

Applauding the "accelerated learning" enabled by teaching machines and pointing to the "almost unanimous" enthusiasm by psychologists, the *Fortune* article again accused teachers of standing in the way of technological progress. "Some teachers oppose the program method because they suspect it might eventually cut the number of teaching jobs: a teacher administering a programed course might be able to supervise three or four times the number of students he could manage in the traditional educational manner," Boehm wrote. "Others argue that the new method 'dehumanizes' education by breaking the personal bond between teacher and student. But what bothers most opponents is that programs seem to them basically more appropriate to an animal psychology laboratory than to a school."[37] The article, which traced the origins of the teaching machine through B. F. Skinner and Sidney Pressey's work, even suggested that it was teachers' reluctance to change their practices and adopt new technology—and not the stock market crash of 1929—that caused the latter's machine to fail.

It's a story that seems to have stuck.

Like most advocates for programmed instruction, Boehm insisted that teachers would be liberated by teaching machines. Liberated, not replaced: "Teachers in schools will be freed from what Skinner calls 'white-collar ditchdigging.' They won't have to dispense routine information or correct home-work assignments. Moreover, if schools are redesigned so that classrooms are broken up into study booths, where each student can pursue the program on his own, the teacher won't have to maintain discipline. . . . The emancipated teachers will have more time for counseling individual students who have problems, and for discussing original ideas that a bright student may propose."[38] Automation would enable the teacher to do *more* work, although that work would change. As Skinner himself put it, "There is no reason why the schoolroom should be any less mechanized than, for example, the kitchen. A country which annually produces millions of refrigerators, dishwashers, automatic washing machines, automatic clothes driers, and automatic garbage disposers can surely afford the equipment necessary to educate its citizens to high standards of competence in the most effective way."[39]

Of course, automation *might* replace the teacher entirely. Skinner and Pressey insisted that was never their intention, but the popular narrative has always floated the possibility that robot teachers are on the horizon.

That was the future depicted in *The Jetsons*, at least. The Hanna-Barbera cartoon appeared on prime-time television during the height of the teaching machine craze. Mrs. Brainmocker, young Elroy Jetson's robot teacher (who, one must presume by her title, was a *married* robot teacher), appeared in just one episode—the very last one of the show's 1960s run, airing on March 3, 1963.

At the Little Dipper School, Elroy confidently talks through the solution to a math problem written on the blackboard. His answer however is gibberish: "And eight trillion to the third power times the nuclear hypotenuse equals the total sum of the trigonomic syndrome divided by the supersonic equation."

"Now one second while I check over your answer," Mrs. Brainmocker responds, rapidly clicking on the panel of buttons on her chest. "Boink!" A slip of paper emerges from the top of her head. "Absolutely correct, Elroy," she reads. "You really know your elementary arithmetic." As she begins to gush about what a pleasure it is to teach students like him, she starts to stutter. "I've got a short in one of my transistors," she apologizes to the class.

Mrs. Brainmocker was obviously more sophisticated than the teaching machines that were peddled to schools and to families at the time. The latter couldn't talk. They couldn't roll around the classroom and hand out report cards. Nevertheless, Mrs. Brainmocker's teaching—that is, her functionality as a teaching machine—is strikingly similar to the devices that were available to the public. Mrs. Brainmocker even looks a bit like the AutoTutor, a machine designed by Norman Crowder and released by U.S. Industries in 1960, which had a series of buttons on its front that the student would click on to input her answers and which dispensed a paper read-out from its top containing her score (see figure 7.2). An updated version of the AutoTutor was displayed at the World's Fair in 1964, one year after *The Jetsons* episode aired.

Teaching machines and robot teachers were part of the 1960s' cultural imaginary. And the desire to replace teachers with robots, actors, operators, and "learning engineers" should not be minimized. But that imaginary—certainly in the case of *The Jetsons*—was, upon close inspection, not always particularly radical or transformative. The students at Little Dipper Elementary still sat in desks in rows. The teacher still stood at the front of the class, punishing students who weren't paying attention. (In this case, that's school bully Kenny Countdown, caught watching the one-millionth episode of *The Flintstones* on his TV watch.) There

Figure 7.2
Promotional photograph of Norman Crowder and the AutoTutor. Rights holder unknown.

were other, more sweeping visions of the future of teaching machines in the late 1950s and early 1960s—Simon Ramo's "A New Technique of Education," certainly. But much of what was touted excitedly as "the future of education" and what was absorbed into the cultural imaginary about that future were very rarely all that different from the present. Looking closely at the technologies in these futures, one finds that they're very rarely all that innovative. Electronic worksheets. Math drills. Televised lectures interspersed with multiple-choice questions.

Proponents of teaching machines in the 1950s and 1960s were quite aware that some of the excitement for their inventions was bound up in the novelty. Students responded enthusiastically to the new devices—but would that last? (A familiar concern to this day.)

Nonetheless the teaching machine—push-button education—was a powerful postwar fantasy. It did not seem to matter that it was a fantasy that failed match the reality of what the machines could do or how often they were actually adopted. Enough excitement was generated in the press and popular culture to pique the interest of a handful of school administrators, willing to experiment with teaching machines in their districts.

8

HOLLINS COLLEGE AND "THE ROANOKE EXPERIMENT"

"Here in the Virginia highlands where the Blue Ridge Mountains meet the Alleghenies," opened a 1960 story widely syndicated by the Associated Press, "an educational revolution is brewing which may sweep the nation."[1] If there was one implementation of teaching machines that was hailed again and again in the press for its "amazing" results[2], it was this one—the project undertaken in the public schools of Roanoke, Virginia.

What was often described as "the Roanoke Experiment" began in early 1960, when Allen Calvin, a psychology professor at nearby Hollins College, received a $68,000 grant from the Carnegie Foundation to assess the potential of programmed instruction and teaching machines—a little over half a million dollars today. In addition to carrying out trials on his college campus, Calvin approached the Roanoke Public Schools—still a racially segregated district at the time, despite the recent *Brown v. Board of Education* decision—about running an experiment there.[3] Conducting research in a school was often politically fraught, as B. F. Skinner had discovered when he tried to convince James Conant to help him make inroads into the schools in Harlem. But the superintendent of the Roanoke district, Edward Rushton, was enthusiastic about new methods and new technologies of teaching, particularly those that would allow the overcrowded district to teach more students.[4]

Calvin worked closely with Rushton to devise the study: ninth-grade algebra would be taught via programmed instruction to eighth-grade students. If the experiment failed somehow and the teaching machines proved ineffective, the students would still be able to take ninth-grade algebra as normal the following year. "Of the 253 eighth grade students enrolled at Woodrow Wilson Junior High School, where the experiment would take place, a third, because of their above-average mathematical ability, had already been selected to begin first-year algebra at the ninth grade level but in conventional classes," Rushton explained. "The students for the pilot study in programmed learning would be drawn, not from these above-average students, but from the other students. If they too could master first year algebra, it would be a clear gain. If not, no one could say that their participation in the experiment had held them back."[5] Rushton made sure that everyone involved was informed and approved of the experiment: the school board, the junior high principal, the teacher, Miss Clintis Mattox, the parents of the thirty-four students selected for the program, as well as the students themselves—on its surface, at least, a rare example of community "buy-in."

The pilot began in February 1960, with the class using Foringer & Company teaching machines—Skinner lamenting, of course, that his Rheem-built machines were not ready. The program was written by Daniel Murphy, a former high school math teacher who was enrolled in the psychology graduate program at Hollins College and who, according to Skinner at least, could barely stay ahead of the students in writing the materials.[6] The students worked on their own and at their own speed. They received no explanations and no help from the teacher; they were given no homework assignments. All the instruction and assessment were done by the teaching machines—each new concept introduced reduced to the smallest possible component and each student learning immediately if his or her answer was right so they could move on to the next step.

According to Rushton, the students who participated in the first-year pilot were incredibly enthusiastic. "They were 'pioneering,' and they knew it," he later boasted.

> As it happened, the class period in which they were to undertake the programmed learning course came immediately after their lunch period. Habitually they rushed through their lunch in order to get to the work. Each day they queued up, impatient to get to their machines and programs. When they got to their desks and began work, they kept at it without interruption, stopping only when the teacher or the class bell forced them to. The result . . . was that they were able to cover in approximately one semester the content of an entire year of ninth grade algebra.[7]

The students' excitement was hailed by the press, and even Skinner remarked upon it, reminiscing in his autobiography about a visit he had taken to the junior high: "The students were at work on the machines when we came in, and when I commented on the fact that they paid no attention to me, Calvin went up to the teacher's platform, jumped in the air, and came down with a loud bang. Not a student looked up."[8]

For Skinner, the Roanoke Experiment was an impressive validation of his ideas.

Rushton and the Hollins College researchers admitted that the students in the Roanoke pilot might have done so well in part because of the "Hawthorne effect"—that is, their performance changed because they knew they were being observed, indeed, that the program they were part of was "pioneering." But they cheered the results of the study nonetheless. At the end of the semester, the students were given a standardized achievement test in algebra that was typically used for ninth-grade students. "What the test seemed to suggest," Rushton wrote, "was that working through programmed materials, without lectures or help from a teacher and without homework, eighth grade students were able to complete a full year's ninth grade algebra course in a semester's time and to score on a standardized test at least 'average' for the ninth grade norm. To say the least, this was highly encouraging."[9] One year later, the students were retested to see how well they could recall what they'd learned on the machines, and, as Rushton gushed, their scores "indicated an average rate of slightly more than 90 per cent!"[10]

The Hollins researchers also asked the students what they thought of their experience, and only five of them said that they preferred traditional teaching methods to programmed instruction. When asked if they would prefer traditional teaching methods, a class taught by machines alone, or one in which the teacher gave help rather than just observed, only two said they'd prefer the first. Eight students said they'd prefer the programmed instruction. The majority of the students—twenty-four of them—said that they would prefer the latter option, using programmed instruction but with a teacher actively taking part in the class.[11] The students preferred working at their own pace, the superintendent insisted, in part because it eliminated the social stigma of excelling or failing at lessons, relating what one student told him: "The eggheads don't get slowed up; the clods don't get showed up."[12]

With such a positive outcome from the pilot program (and with hundreds of headlines lauding the school district's innovation), Rushton agreed not only to continue the Roanoke Experiment but to expand it substantially for the 1960–1961 school year. All three of the city's high schools would be involved (including the Lucy Addison High School, which served the Black high school student population)—in total, eleven teachers, thirty-two classes, and approximately nine hundred students across three courses, Algebra I, Algebra II, and Plane Geometry.[13] The Foringer teaching machines were replaced by programmed textbooks as the devices were simply too expensive to purchase in great numbers—the machines cost $70 apiece, while the textbooks were only $12.[14] There might be even more cost savings, Rushton claimed, if students were able to reuse some of the programmed textbooks.

Students were randomly assigned to one of three kinds of classes: the first was a "conventional" class, in which a teacher taught using traditional textbooks in the traditional manner. The second kind of class, the "help" class, used only programmed materials, as the initial pilot study had done, but this time a teacher would be there to "assist the student in any way he saw fit."[15] The third class type was the "no help" class, which also used programmed materials but in which the teacher "was not permitted to give the student any assistance in learning the material other than to discuss with him the results of periodic course examinations."[16] The students in the programmed instruction classes were not allowed to take their programmed textbooks home, and they were not given homework, while the class taught in the traditional way received homework assignments as usual.

A number of problems occurred early in this second year that made what followed quite different from the successful pilot program. The programmed textbooks were late arriving in Roanoke, Rushton reported. "Not until the last minute were we sure they would arrive in time at all. The result was that there was no time to follow the procedures which had worked out so well for the pilot study the previous school year. We were unable, as before, to call in students and parents, to explain the experiment, and to ask their cooperation, nor were the programs available for teachers to examine."[17] No surprise, then, that parents had a lot of questions: Why is there no homework? Why is my child in a class without a teacher? Is my child learning enough? Will courses taught by programmed instruction be accepted for college admissions requirements? Will the Virginia Department of Education grant a full year's credit for a course that my child finishes in much less time? Rushton and his team tried to respond to all of these and assure parents that students would benefit greatly from the new teaching technology. But one question was (and still is) difficult to answer: Why is my child being used as a 'guinea pig' for an experiment?[18]

Despite all these concerns, Rushton maintained that the use of programmed instruction during the 1960–1961 school year was, like the first year's trial, largely a success. As the students in the programmed instruction courses were allowed to move at their own pace, "some of the students finished an entire year's work by December, most of them finished before the end of the term, and a few required more than an academic year to complete the programmed course. The students who finished early were encouraged—but not required—to use the remaining time in undertaking advanced mathematics courses. Some did so, while others chose to devote the time to studying for other subjects," Rushton reported.[19] Seventy-two percent of the students in the "help" class and 77 percent of the students in the "no help" class in Algebra I finished the course early. Fewer students failed (and failed to complete) the programmed courses than did students in the traditionally taught ones. At the end of the year, when standardized achievement tests were administered to the students, the researchers found that the mean score for those in the "no help" method was significantly higher than those in the conventional classrooms. (The difference between the scores of students in the two types of programmed classes was not significant.)[20]

But the results in the other courses were not as encouraging—nor were they much mentioned by Rushton in his chronicle of the Roanoke Experiment. The students in the conventional Algebra II class did better than those in the programmed ones, and there was no significant difference between the student performance in any of the Plane Geometry classes.[21]

The results then were hardly overwhelming. Nonetheless, the district opted to extend its use of programmed instruction, expanding it to other subject areas. The student and teacher reactions were enthusiastic enough to justify continuation. According to Rushton, teachers reported that programmed instruction had changed their own pedagogical practices, making them rethink how they taught and what they assumed certain students could be capable of. And as he described it, at least, any of the struggles that the teachers might have faced in implementing a new curriculum and new instructional practice did not dampen their attitudes toward teaching machines. Rushton did admit that teachers should have been involved much earlier in the planning process, estimating that it took as long as four months for some of them to feel confident with the new technology.[22]

Rushton was certain that

> Programmed instruction, through the use of teaching machines and programmed textbooks, has proven effective in providing for student achievement in and retention of high school mathematics. Test results indicated that student achievement was highly satisfactory in the learning of factual knowledge of high school mathematics by this method of instruction (as measured by the standardized tests); and no significant difference in retention resulted when students were retested from four to eight months after completion of courses in high school mathematics.[23]

There were "revolutionary implications"[24] too, he contended, for adult education, homeschooling, and curriculum expansion. Programmed instruction would change how teachers taught—focusing less on "imparting factual content and correcting homework assignments," and instead be free "to counsel individual students, to discuss problems with them in small and large groups, to help slow students, to explore original ideas with bright students, and to do creative planning."[25] Echoing the Sputnik-era language of crisis, Rushton wrote that "at a time when our country's schools are face to face with an educational challenge of staggering proportions, [programmed instruction] possesses vast potential for teaching many more students than had been possible before with a gain in the quality of learning."[26]

The benefits of programmed instruction might have been far less transformative than Rushton liked to describe, but the pilot program gave the district an incredible boost of positive publicity during a period when the city was struggling to desegregate its schools while maintaining its reputation for "civility" and progressivism.[27] Although Rushton (and the press) would frequently cite the speed with which students moved through the programmed materials, no definitive conclusions could really be drawn from the research in the schools, in part because of the design of the study itself. There were a number of uncontrolled variables, for example, including a teacher who failed 30 percent of her students and who was described as "hostile" to the project.[28] More important, perhaps, the connections between the teaching machine industry and the Roanoke researchers were never really full disclosed or interrogated.

Since May 1960, the first year of the Roanoke Experiment, Encyclopedia Britannica Films (EBF) had been deeply involved in the project, publishing some of the programming materials, tests, and instructional guides that Hollins College had initially developed under the TEMAC brand name—short for TEaching MAChines. Rushton, who along with Allen Calvin retained a tight control over the narrative in the press, wrote two books touting the project—the first in 1963 and the second in 1965—both published by the encyclopedia maker.

In January 1960, John Everett, the president of Hollins College, left his position to become a consultant at EBF—just one of many, many Hollins College staff involved in the teaching machine initiative who later joined the company. EBF offered to fund Allen Calvin's research on programmed instruction and create a Center for Learning and Motivational Research at the college. But in December 1960, the college's board of trustees rejected the plan, telling the press they did not think "the latest EBF proposal gives them sufficient control over both the personnel doing the research and the kind of research done."[29]

Trouble had been brewing for some time at Hollins College with regard to the programmed instruction research there. Calvin, along with his colleague Maurice Sullivan, a professor of modern languages, had developed other instructional materials, in addition to the math programs piloted in the Roanoke secondary schools. Sullivan was particularly interested in reading instruction, eventually publishing a series of programmed books called *I Can Read* and one called *Reading Readiness*.[30] Sullivan and Calvin, along with Cynthia Buchanan, a former student of Sullivan's who had returned to Hollins College as a language instructor, received numerous grants to develop their programmed materials—from the Carnegie Foundation and the US Office of Education, for example—and had rapidly expanded the number of staff and students working on their research.

Many of these materials followed what they called "the Skinnerian program block," each frame a small step forward in content and "a totally independent unit."[31] But Sullivan began to feel as though this method of programming was too redundant, too boring, particularly for struggling readers, and he developed a technique he called "chaining," so that each frame was better connected to the previous one without so much repetition. The researchers at Hollins College also quickly discovered that preparing their materials for a machine—be it an audio recorder or a film-based machine or a paper-tape system—was too cumbersome. The available machines were all too expensive, and the constant revisions to the programming materials—something that Susan Meyer Markle had touted as a key benefit to this new instructional technology—would mean that new filmstrips or tapes had to be produced continually. Teaching machines "only enriched the people who made machines," Sullivan complained; so, the Hollins team revised all their materials to work without them.[32]

The Carnegie grant had helped expand the size of the research group at Hollins significantly, but a subsequent million-dollar investment from Encyclopedia Britannica Films in 1960 enabled Sullivan, Calvin, and Buchanan to hire more than seven hundred people—a much, much larger team than any other university's lab (including, of course, Skinner's at Harvard).[33] This staffing level was an indication that the programmed instruction work might have been "by far the largest and most important project at Hollins College, which had less than 1,000 students of its own." As Sullivan's company would later tell the story, "Every year, Sullivan received more grant money for his own projects than the rest of the college had received in its whole history. Sullivan's assistant programmers made more money than full professors outside his group did. . . . Understandably, relations between Hollins College and Sullivan's staff were strained."[34]

After the board of trustees at Hollins College blocked the plan for EBF to fund an expansion of the programmed instruction research facilities at the school, Sullivan and Calvin relocated their team to Palo Alto, California, where EBF opened a new office, The Britannica Center for Studies in Learning and Motivation. The center opened in September 1961 but closed just three months later as the EBF business staff and the former Hollins College researchers were unable to work together—one of a series of clashes between the academic and the commercial orientations of the teaching machine movement.

9

TEACHING MACHINES INC.

By 1962, nearly two hundred companies were producing, or were planning to produce, teaching machines or programmed books, "for schools, industry, the armed forces, government agencies, and the home"—that is, at least, according to *Changing Times*, which estimated that sales would surpass $100 million by the end of the decade.[1] (That's about $850 million in today's dollars.) Another article, published in the *Reporter* the same year, was more conservative with its claims: it put the number of companies in the teaching machine business at sixty—companies it said were selling their devices "at prices ranging from less than five dollars to more than $2500, devices with names like Learn Ease, Edumator, Ed-U-Data, Visitutor, Didak, Tutor-Matic, Instructon, Redi-Tutor, and Omnibox Teachall."[2]

No matter the number—both sixty and two hundred are impressive considering widespread publicity about teaching machines had commenced just a few years earlier—the competition meant that these companies had to rely on a variety of sales tactics in order to differentiate themselves and attract customers. "One firm, the Univox Institute, is offering teaching machines through supermarkets and by mail," wrote Spencer Klaw in the *Reporter*. "In observance of the old rule that the real money is in selling the blades, not the razor, Univox does not charge mail-order customers for its machine. Anyone who would like to try its 'Fabulous New Teaching Machine Auto-Mated Speed Learning Method' is invited to send in $14.95 for a program in, say, grammar or selling, with which the company will throw in free of charge a 'personal leatherized Univox teaching machine.'"[3] Teaching machines were heavily promoted in magazines and newspapers, with promises that learning would be faster, easier, and more fun. "Guaranteed to Improve Your Child's School Marks," read one ad in the *New York Times*. "New Automated HONOR Teaching Machine Helps Your Child Learn Faster, Do Better in ALL School Subjects," boasted another.[4]

The sales pitches weren't just in print; teaching machine vendors were, quite literally, knocking on people's doors. Arguably the most successful teaching machine company, Teaching Machines Inc. (TMI), attained this status through a relationship with the publisher Grolier, whose army of door-to-door encyclopedia salespeople were enlisted to hawk the device.

Teaching Machines Inc. was cofounded in 1959 by University of Pittsburgh professors Lloyd Homme and James Evans, two years after Homme returned to his home campus from his stint in B. F. Skinner's teaching machine lab at Harvard.[5] The two also invited Robert Glaser, another colleague from the University of Pittsburgh (and, in 1960, the coeditor, along with A. A. Lumsdaine, of *Teaching Machines and Programmed Instruction: A Source Book*) to join the company. They also brought on board Ben Wyckoff, whom Homme had known from graduate school at Indiana University.

Originally headquartered in Pittsburgh, the company relocated to Albuquerque in 1960. In November of that year, TMI received investment from Grolier, one of the largest encyclopedia publishers in the world. As part of the deal, Grolier agreed to use its 5,000-person door-to-door sales force to sell one of TMI's teaching machines, the Min/Max, alongside its *Book of Knowledge* and *Grolier Encyclopedia* sets: buy an encyclopedia; get a teaching machine for free. The stock market certainly saw this as a smart move. Upon news of its investment in the teaching machine business, Grolier's share price jumped from $32 to $50.[6]

The Min/Max teaching machine was designed by Dudley Cornell; its name was short for "minimum time, maximum learning." The original version was made of metal and was slightly larger than a typewriter. The Min/Max could hold several pages of programs at once, and the student would move through the materials by looking through a small window to see a question, writing the answer in another window and then checking to see if he'd gotten the answer correct. It was hardly "mechanical" at all, as the student had to push the paper forward to advance to the next question. The Min/Max machine soon was redesigned to be smaller, lighter-weight, and ostensibly more portable.

The updated Min/Max was about eighteen inches long and ten inches across—dimensions that still rendered it bulky, even though it was not heavy since it was made of plastic. The teaching machine's proportions made it too big to rest comfortably in one's lap, particularly a child's, and it took up a lot of space when it perched on a desk or table. The only mechanical part of the machine was a dial on each side used to advance the paper-based programming materials with a roller mechanism similar to that of a typewriter. The lid lifted to insert those pages—no more than 100 sheets at a time, the instructions cautioned (see figure 9.1).

One representative course, TMI-801 Fundamentals of Electricity, contained 150 sheets of paper, printed on both sides, each side with around five or six "frames" of instructional content. The student would slide about half the papers

Figure 9.1
Image of the TMI teaching machine and programming materials used with permission from The Strong National Museum of Play, Rochester, NY.

into the machine and spin the knob until the first question appeared in the clear plastic window at the top. The first few frames introduced the student to programmed instruction, explaining how to read the question, write the answer in the blank space given, then push the paper up until the answer could be checked. "The steps in this program are fixed so that you should be right most of the time," the instructions explained. The course then offered a couple of sample questions that demonstrated how it worked: "If the answer to a question is a word to be filled in the blank, it is shown with a line like this _____. George Washington was the first _____ of the U.S."

The first electricity question: "All matter is made of molecules (say MAHL-e-kules). Wood is made of molecules. Water is made of _____." (The answer: molecules.) The course moved slowly from there—those incremental advances Skinner had advised would minimize errors and therefore serve as a positive reinforcement. It isn't until question 9 that the student would get to atoms, question 17 that they'd learn about electrons. It was easy to get bored well before then. Even with the promise that the student could "move at their own pace," that pace was necessarily slowed by the small, incremental steps in each frame. Moving on from electrons, the student would still have had 1,436 frames to go.

Even though several members of the TMI team had worked with Skinner and their machines drew on his ideas, he was not particularly supportive of their endeavors. Soon after the TMI-Grolier partnership was inked, he wrote to Theodore

Waller, a member of the board of directors of Grolier, relaying concerns that "a friend of mine on the West Coast" had about the methods of the publisher's encyclopedia sales force: "The Grolier Society is peddling their machines and courses with your name attached," Skinner's friend had warned him. "We were visited by a house-to-house salesman, who wanted us to buy a package deal for the education of our children. This consisted of a machine and four programs for $75. The ring of charlatanism was in every syllable, especially given what I have picked up from other sources about the Grolier programming. Do you know what these people are up to, and whether they have a decent respect for the opinions of honest scientists?"[7] Skinner didn't add any commentary, simply transcribing this excerpt from his friend's letter, telling Waller that he just "thought you might want to know."

It was hardly the first time that someone had tried to use Skinner's name without his permission to promote teaching machines—and it wouldn't be the last.[8] Although Skinner had signed a contract with Rheem Manufacturing Company to develop a teaching machine of his design in 1959, he had little faith that that corporation would properly protect his name and reputation. He decided to take the matter into his own letter-writing hands—at least as far as some of the perceived improprieties of Grolier and TMI were concerned.[9] He wrote to his former colleague Lloyd Homme to complain, and he wrote again to Theodore Waller, who assured Skinner that he would summon the California sales manager to New York "to be put on the carpet."[10] Waller promised that Grolier would make every effort to make sure its salespeople did not mention Harvard and did not use the famed psychologist's name.

Skinner wasn't the only person to bristle at the encyclopedia salespeople's behavior. "In Middlesex County, Massachusetts," one article reported, "when salesmen first began selling the Min/Max, the county PTA Council publicly warned parents to be on their guard. A salesman may imply, for example, that he has been sent around by Johnny's teacher—whose name he may have obtained along with the names of Johnny's parents, in return for giving Johnny's school a free set of *The Book of Knowledge*." The article decried these sales tactics, adding, "Even when tricks of this kind are not used, many educators object strongly to Grolier's methods on the reasonable ground that an encyclopedia salesman, who may have just switched from selling storm windows or secondhand cars, is poorly qualified to prescribe for a child's educational needs."[11]

Of course, it wasn't just the salespeople of Grolier who were viewed unfavorably. One of the publisher's main competitors, Encyclopedia Britannica, had been ordered in 1961 by the FTC to stop its deceptive sales practices and pricing schemes.[12] Yet despite the bad impression that the door-to-door salespeople might have left in many people's minds, encyclopedias were still viewed as a worthwhile expenditure. Some of this popularity, a 1945 profile of Encyclopedia Britannica in the *Saturday Evening Post* contended, was merely "to keep up or forge ahead of the Joneses."[13] Possessing a set of leather-bound encyclopedias was a sign of prosperity.

It was a sign too of a curiosity about the world and a commitment to knowledge—so much so that ownership of a set of encyclopedias was one of the factors examined in *Equality of Educational Opportunity*, better known as the "Coleman

Report" and arguably one of the most important education studies of the twentieth century.

As part of the passage of the Civil Rights Act in 1964, Johns Hopkins University researcher James Coleman was commissioned to study the effects of race on inequality in the US public school system. In a massive sociological undertaking, Coleman and his team collected data from 4,000 schools, 66,000 teachers, and almost six hundred thousand first-, third-, sixth-, ninth-, and twelfth-graders.[14] Coleman found that public schools were—no surprise—highly segregated throughout all parts of the country, not only in the South, and there were significant gaps in achievement between white and Black students. But Coleman's report argued that inequalities in school resources alone were not sufficient to explain the differences in student achievement. Instead, he argued that family background—parental income, education, and aspirations for children—had a strong influence on students' test scores. He observed, "one implication stands out above all: that schools bring little influence to bear on a child's achievement that is independent of his background and general social context; and that this very lack of an independent effect means that the inequalities imposed on children by their home, neighborhood, and peer environment are carried along to become the inequalities with which they confront adult life at the end of school."[15]

Coleman's study found that "the average white high school student attends a school in which eighty-two percent of his classmates report there are encyclopedias in their home."[16] The number was far lower for Black students (and for Puerto Rican students as well)—only half of Black elementary school-age children went to school with classmates who had encyclopedias at home.[17]

Buying an encyclopedia set was a significant investment, and payments were typically made in installments. A set of twenty volumes of *The Book of Knowledge* cost $149.50 in 1962 (about $1,200 in today's dollars); a set of twenty-four volumes of the *Encyclopaedia Britannica* ran from $398 to $549 (between $3,300 to $4,600 today).[18] The median household income in 1962 was about $6,000—less than half of that for Black families in the South.[19] Purchasing an encyclopedia was a major investment for families of all backgrounds. It was an aspirational gesture. The books might never be referenced or read. But their presence in the home was a visible symbol that education mattered.

And perhaps the same case could be made for teaching machines, although the price offered by those door-to-door salespeople was far less steep: $20 to $25 for a machine and $7.50 to $15 for a program.

In just the first two years of their partnership, TMI and Grolier sold over 150,000 teaching machines.

The Min/Max was not the only, or even the first of TMI's teaching machines. Ben Wyckoff, a former student of Skinner's at Indiana University who'd also come to Harvard for a summer (in 1950) to work in the pigeon lab, had developed the company's first machine, one that was far more sophisticated than the rather cheap and crude Min/Max.[20]

The initial version of Wyckoff's machine was designed to teach reading. It displayed lessons on a small screen and was operated by pressing its five keys. The slides that were projected onto the screen displayed words that were all missing letters. The student was supposed to complete the words by pushing combinations of the keys in order to supply the right missing letter. When these keystrokes were correct, the film would advance to the next slide, displaying the completed word on the screen. The machine, the Wyckoff film-tutor as it was called, had a hefty price tag of $445—about $,3800 in today's dollars.[21] The price wasn't the only drawback, as Wyckoff himself admitted in the patent application he filed in 1960: "the student must either first learn or have available a conversion table for the interrelationship between the keys to express a specific letter."[22] (Wyckoff would later add a full keyboard.)

Even with these shortcomings, the Wyckoff film-tutor was superior to the Min/Max in several ways. The film-tutor did not use paper, for starters. And the Min/Max used *a lot* of paper. According to a presentation delivered at the National Society for Programmed Instruction's conference in 1962, C. J. Donnelly, the vice president of Grolier's teaching materials division, confessed that "TMI-Grolier alone, in the printing of programs, used 1000 tons of paper in 1962," which was why, he said, the company was exploring film-based machines. "We also accounted for something like 500 tons of plastic for the relatively uncomplicated, manually-operated, auto-instructional device we call the Min/Max," he added.[23] The second benefit to the Wyckoff film-tutor was that it allowed for the display of question items of various lengths—items much longer and more intricate than the short Q&A that fit on the Min/Max screen (or on the devices that Skinner designed)—a feature that A. A. Lumsdaine praised in one of the articles he contributed to his and

Glaser's "source book." It was a feature that the film-tutor shared with Norman Crowder's AutoTutor.[24]

Despite these attributes, Grolier decided it did not want to distribute the Wyckoff film-tutor for TMI. It was, after all, twenty times the price of the Min/Max and three times the weight—entirely unsuitable for door-to-door sales. And as the press and Skinner had rightly observed, the Grolier sales force knew little about why the bigger, heavier, and costlier machine might be better designed. It would be an almost impossible sell. And without the help of Grolier, the Wyckoff film-tutor would never be a viable commercial product, let alone a commercial success.

Although the encyclopedia publisher wouldn't distribute the machine, it was more than willing to distribute the program Wyckoff had designed for it—a program that would teach children to read. In his 1962 book on teaching machines, Benjamin Fine, the former education editor of the *New York Times*, gushed about the Wyckoff device and the work in early childhood education undertaken at TMI's headquarters in New Mexico: "Why can't Johnny read?" he asked, echoing the title of the 1955 bestseller on phonics by Rudolf Flesh.[25] But Fine had a different answer than Flesh did. It wasn't a lack of phonics instruction. Johnny couldn't read, Fine declared, "because he never took the Wyckoff teaching-machine program!"[26]

In 1962, TMI added another device to its catalog: a version of B. F. Skinner's "air crib." Skinner had attempted to restart production of his earlier invention a few years prior, through work with John Gray, an entrepreneur who'd founded the Aircrib Corporation in 1957, a move that had prompted another flurry of publicity—good and bad—about "babies in boxes." Skinner was "not much impressed" with TMI's version "for several reasons."[27] He later related in his autobiography that

> Lloyd Homme's company, Teaching Machines Incorporated, produced a model in which most of the cabinet work was replaced with a great plastic bubble. Gray agreed to license the name "Aircrib" if the model had our approval. It seemed safe enough, but I did not like the resonating acoustics of a bubble or the radiant heat loss through clear plastic. A revised model, attractively designed, had fewer of these faults. I also told Lloyd about the musical toilet-training seat I had designed for Deborah, and he thought they might add it to their line. But his company was soon taken over by the Westinghouse Corporation, which was interested only in teaching machines.[28]

Skinner didn't quite get the story right. (Skinner is wrong in several places in his autobiography in a similar sort of way—wrong about names and dates and other details. Such are the easy and often unintentional errors one makes in a memoir, no doubt. Nevertheless, it is worth thinking about what it might mean to have as important a figure as Skinner, someone who has been granted the power to tell the history of education psychology and education technology, be an unreliable narrator.) It was actually a spin-off of Teaching Machines Inc., the TMI Institute, that was acquired by Westinghouse in 1965. Lloyd Homme was chosen to head the new Behavior Systems Division; Donald Tosti was hired as his assistant.[29] Robert Glaser had already departed TMI in 1963, and Ben Wyckoff had stepped down as chairman of its board shortly after that (to pursue, it's worth noting, more corporate-oriented programmed instruction that Grolier was not interested in publishing).[30] In 1966, the remaining founding members, James Evans and Dudley Cornell, who'd long struggled to keep TMI operational, declared bankruptcy. TMI's assets were acquired by Grolier for $633,000 (about $5 million in today's dollars), a sum that "will allow payment of more than 95 cents on the dollar to TMI creditors"—"one of the highest-paying bankruptcies" he'd ever seen, the company's lawyer boasted to the *Albuquerque Journal*.[31]

The end of Teaching Machines Inc. was more of a slow dissolution than a sudden takeover. But even with its closure, it's hard to label it a complete failure. Over the course of the company's existence, TMI had sold almost a million Min/Max machines and more than two million programs[32]—a figure that had surely made the teaching machine business appealing to some of Grolier's major competitors in encyclopedia and educational publishing. Field Enterprises Educational Corporation, publisher of the *World Book Encyclopedia*, for example, offered a "combined teaching machine and family quiz game."[33] And Crowell-Collier explored selling a Collier's Home Teaching Machine alongside its encyclopedia. The most high-profile competition was that of Encyclopedia Britannica Films, which had signed a contract with Hollins College's Allen Calvin in May 1960. The *New York Times* announced that deal in June: "Teacher Machine to Be Ready in '61," a headline that surely irritated B. F. Skinner as those were the exact same words that the newspaper had used to describe his Rheem-built machine the previous year—a machine that still was not ready.[34]

In that 1959 *New York Times* article, Skinner had stated that the only real obstacle standing in the way of widespread adoption of teaching machines was the lack of good programs. The interest of encyclopedia publishers in teaching machines—and more importantly in teaching machine *programs*—had helped to address the availability of programmed instruction. By 1963, 188 of the leading publishers in the United States had programmed materials for sale.[35] (Although in Skinner's estimation, none of this had really addressed the lack of *good* programs.) As with the textbook and trade publishers who had jumped on the teaching machine bandwagon—Harcourt Brace and Doubleday, for example—programmed instruction enabled these encyclopedia publishers to appear innovative and attuned to the latest in educational science.

The availability of programmed instruction materials also helped publishers appear responsive to a changing economy and to changing expectations of education's role in it. Buying an encyclopedia might have been a symbolic gesture that knowledge mattered. But there were practical reasons too to invest in study at home, and there was a thriving industry of self-help, self-improvement, and self-education materials that were aimed at adults, not just at children. Programmed instruction and teaching machines were another way that these lessons could be delivered—delivered with promises that, by using them, new skills could be more rapidly attained.

The military had long used teaching machines to train its members, and corporate job training was seen as another potential market for teaching machines—companies like Hughes Aircraft, for example, invested heavily in them for a time. But not all of this training was to be underwritten by employers. At least that was the message of magazine and newspaper advertisements that encouraged men (yes, mostly men) to buy programmed instruction materials and teaching machines in order to learn new technical skills and join some of the booming postwar industries.

Although the founders of TMI—Lloyd Homme, Robert Glaser, Ben Wyckoff—tried in their work to strike a balance between designing programmed materials for use at home and at work, it may be that Norman Crowder best exemplifies the attempt to address the multiple markets for job training. His "scrambled textbooks" sold for less than $4. The AutoTutor, manufactured by his employer Western Design, a major manufacturer of industrial products, was priced at $1,500—about $13,000 in today's dollars. Someone could use Crowder's paper-based teaching machine to learn electronics at home or use the expensive equipment to learn electronics at work.

The growing markets for teaching machines represented a "learning boom," as the rapid changes in technology would soon force everyone to constantly train and retrain—"'ability to learn' would gradually replace 'ability to do the job,' one reporter predicted."[36] Teaching machines were sold with the promise they would aid with precisely that. As one ad for the Min/Max put it: "Self-improvement is key to success in today's demanding Space Age."[37]

10

B. F. SKINNER'S DISILLUSIONMENT

In June 1960, Rheem's general attorney, Walter Lewis, wrote to B. F. Skinner, concerned that Encyclopedia Britannica might be infringing on his (or rather, Rheem's) patent. Lewis enclosed an article from the *New York Times*, "Teacher Machine to Be Ready in '61"—the same prediction and almost identical headline that the newspaper had given its coverage of the Rheem-Skinner machine the year before.[1] "Perhaps you should have your patent lawyers examine the point more closely," Skinner retorted.[2]

Skinner was quite aware of the competition that other teaching machine companies posed. Eastman Kodak, for example, had a new film-based device he thought interesting although he wasn't convinced "the art of film presentation for individual use is anywhere near what it needs to be to make a film machine feasible at this time."[3] But he was increasingly concerned that many schools were opting to pursue programmed instruction *without* teaching machines. He'd heard from several of the school officials who'd visited

Roanoke that they'd been given a rousing sales pitch "in favor of workbooks" by the Encyclopedia Britannica salespeople, something that convinced Skinner that Rheem needed to do more to improve the availability of programs that would work on the devices it was building.[4] Despite having previously admonished Rheem for using his name to market its products, Skinner encouraged the company to better leverage their relationship in order to stay ahead of the competition.

In February 1961, the revised patent for Skinner's Teaching and Testing Machine was filed with the US Patent Office.[5] But there was still no Skinner device ready for sale.

In early April, Skinner penned what he later described as a "strong letter" to Rheem's executive vice president, C. V. Coons, listing his "anxieties" about the status of the teaching machine effort at the company. Skinner was worried that "properly designed machines are not yet available," and he detailed the problems with the three machines that the contract between the two parties governed, as well as a fourth machine under development.[6] Skinner claimed that his feedback had been ignored, and that machines had been shipped out for public testing despite his protests about their readiness.

Skinner was also upset that Rheem was not doing enough in his eyes to acquire the licensing to programming materials. He explained to Coons the attempts that he'd made to try to get Rheem to work with Allen Calvin,

> but he was unable to get so much as an answer to his inquiries and, instead, bought 80 machines from the Foringer Company. When bugs developed, he turned to a programed

> textbook system. I re-established good relations with him, and he assured me that all the programs he prepared would be available for the Rheem machine. I urged the Rheem people to develop this contact, but nothing was done. For a while it was argued that it would be ill-advised to contact Calvin, who had now associated himself with Encyclopedia Britannica Films, while trying to negotiate with Grolier. Nothing has come of the Grolier association, and I now understand that the Britannica Films are developing a machine with McGraw-Edison. They are putting two million dollars into the development of programs, which are not available to Rheem.

Skinner chastised Rheem with failing to build relationships with publishers, but at the end of the day, he argued, "Rheem does not need to pay for the production of programs. It needs only to make available good machines."

Skinner continued his list of complains, his frustration building from page to page in his letter to Coons. "Rheem apparently has no long-term plan," he wrote in dismay. The company's executives, Skinner charged, seem to lack faith in the Didak 501, despite the resources being put into that model's development. In summary, "Rheem has not yet produced a teaching machine which meets with my approval. . . . I am glad to have your assurance of Rheem's continuing interest, but if we are to go on working together, action on a much more extensive scale is called for." Then Skinner turned to what would become his new tactic for dealing with the company: threatening to dissolve their contract. "I have already lost two years of valuable time in embarking on this development, and I do not think the Rheem Company should try to hold me to our contract unless they are willing to take immediate action on an appropriate scale."

Coons wrote back a week later, acknowledging the letter and promising he was "investigating the entire matter."[7] It took almost a full month, however, for him to respond to Skinner's charges in detail. "Our basic objective continues to be one of maintaining a position of world leadership in the profitable design, production and distribution of automated teaching system."[8] Coons informed Skinner that the company would be establishing a Corporate Teaching Machine Design and Engineering Group—"actually a continuation of work we have had under way since 1959." Coons assured Skinner that the company's plans included the continued development of the Didak 501, a complete redesign of the Didak 101 "preverbal" machine, and continued investigations into the possibility for keyboard-input and rhythm-teaching machines. Coons also said that Rheem was considering forming a subsidiary corporation "engaged exclusively in program development, publication and distribution." Skinner was pleased to hear these reassurances, but he still wanted more details.[9] What *exactly* were the plans for the preverbal machine? Would his input on the 501 finally be taken seriously?

On June 2, 1961, Coons wrote to Skinner again, stating that Rheem wanted to exercise its option to extend their July 1959 agreement.[10] The letter specified the particular products that fell within the contract's definition of "teaching machines" and would therefore be subject to royalties: the Didak 101 multiple-choice machine, the Didak 501 write-in machine, and the Didak 701 recall-key machine. Devices that were excluded from the definition of teaching machines in this agreement were Rheem Califone's lines of school phonographs, tape recorders, language laboratories, and radios, as well as the "Solartron-Rheem psychomotor skill trainers, including the Keypunch Trainer, the 10-key Adding Machine Trainer, and the Typing Trainer," the latter devices invented by cybernetician Gordon Pask.[11]

When Skinner's attorney Donald Rivkin reviewed the letter, this delineation of what was and was not contractually a "Skinnerian" teaching machine gave him pause. "As you know," Rivkin explained, "a machine other than the three types specified must, in order to qualify for royalty payments, satisfy two requirements; it must be one which is (a) 'developed in coordination with Skinner under this agreement,' and (b) which is 'based upon or utilizes the procedures, techniques, research findings and scholarship which Skinner has developed or performed or which he may hereafter develop or perform.'"[12] Rivkin felt as though Rheem should also have to pay royalties on machines Skinner helped the company refine. Even if these royalty issues were addressed satisfactorily, Rivkin agreed with Skinner that Rheem had clearly defaulted on several key elements of the existing agreement. He suggested that Skinner itemize some of the company's specific failures so that any claims it might make about its "best efforts" could be more easily challenged legally, should the need arise.

After the coaching by his attorney, Skinner responded to Coons, writing that

> I assume that the letter is in no way intended to alter the terms of our agreement. I do not expect to receive royalties on the key-trainers or the audio devices manufactured by the Califone Company so long as these do not incorporate features worked out in consultation with me. On several occasions during the last year and a half, however, I have recommended that Rheem design and supply a different type

of audio-device to be used with teaching machines in a new kind of language lab. If Rheem eventually decided to do so, such a device would, of course, fall within the scope of our royalty agreements.[13]

Skinner also pointed out that, while he appreciated receiving encouraging news from the company about its plans for Skinner's teaching machines, "exactly the same assurances were given to me in our meeting on January 4 in Mr. Walker's office but . . . during the ensuing six months no action has been taken commensurate with these assurances." It had been two years since Rheem and Skinner signed their initial agreement, Skinner reminded Coons, "and Rheem has not yet produced an acceptable machine of any kind."

Skinner informed Rheem that he could not sit idle as the company continued to, in his estimation at least, default on their agreement. He demanded that Rheem complete the manufacture of 200 Didak 501 machines redesigned to his specification and that the company complete the long-delayed overhaul to the Didak 101. If the company did not act, Skinner again threatened to terminate the agreement. "It is my hope that we can continue to work together. But I can no longer overlook the fact that I have been forced to remain inactive during important early years of a world-wide movement for which I myself am largely responsible, and I do not feel that I can continue to do so, both for personal reasons and because of my interest in seeing the teaching machine movement progress as rapidly as possible in the right direction," he concluded.

Coons wrote back to Skinner a few days later, disturbed by the accusation that Rheem might be in default of the January 1959 contract. "We do not believe this is a proper and fair conclusion."[14] Coons insisted that Rheem had consistently maintained its interest and activity in the field. He pointed out that, while Rheem had originally agreed to spend $224,000 on developing the machines, it had in fact spent $300,000—an indication, one might suppose, of Rheem's serious commitment, rather than Skinner's interpretation of the figure, which was that the company was obviously just bumbling along. Coons reminded Skinner that it would never have been prudent for Rheem to begin manufacturing teaching machines without having some assurance there was in fact a market for the devices. Coons suggested that Skinner had misinterpreted the original agreement. Rheem had never *promised* it would produce machines on a mass scale. Coons apologized that Skinner felt he'd been forced to be inactive in the field and wrote that "while we believe a cessation of our relationship would not be in either of our best interests, it would appear you might prefer to end the limitations on your services to others in regard to the teaching machine field." In other words, Skinner was welcome to go find another company to work for or to consult with.

Skinner remained certain that there was a giant market for teaching machines—a market that could include every classroom in the country, perhaps in the world—and that he could provide the scientific leadership to help usher in an era of programmed instruction. But this was not a business certainty, and he recognized that. "I am not interested in business, have never consider[ed] giving much time to it, let alone all my time," Skinner recalled, "and am willing to settle for the satisfaction of having started it all—that is, the teaching machine movement."[15]

In early September, Skinner wrote to Coons again, challenging many of the assertions the Rheem vice president had made in his earlier letter. Contrary to Coons's claim that Rheem had remained engaged in the teaching machine field, Skinner charged that as far as he knew "there were only two brief periods of substantial activity" by the company, including the few months in the summer of 1959 when the Didak 101 and Didak 501 were produced for display at the American Psychological Association conference in Cincinnati.[16] Changes to Skinner's original design—changes suggested by another psychologist that were made without Skinner's knowledge—had been eventually corrected, but Skinner claimed that none of the other alterations he'd suggested for the Didak 501 had ever been implemented until it was time for the device to be entirely revamped. "This was admittedly an expensive operation in which dies and other equipment for mass production were secured," Skinner recognized, but that was all the more reason for letting him weigh in on the design before that stage. "Nothing further was done until the summer of this year, as I discovered on my recent trip to the West Coast," Skinner continued. "I think you will find that a very large part of the $300,000 went to market research, promotional activities, demonstrations of machines, field tests of machines, and general problems of management which were not specified in our agreement."

Skinner went on to complain about other things, including the number of names that he'd forwarded to Rheem of people interested in buying or experimenting with the machines, many of whom had never been contacted. "Where other psychologists have been able to extend their activities in association with other companies, my own activity has mainly taken the form of trying to postpone activity until machines might be made available," he lamented. Skinner closed his letter with a few conciliatory words, admitting that he was "impressed" with Rheem's renewed activity.

That included attempts to gain the publication rights to teaching machine programs owned by Appleton-Century-Crofts Inc.[17] Under the proposed deal, the publisher would prepare some of its materials for use with the Didak 501. But when Rheem demanded *exclusive* rights to the programming, Appleton-Century-Crofts balked.[18] Shortly after that deal fell apart, Rheem suffered another blow to its attempts to secure programs for its teaching machines. McGraw-Hill informed the company that it would not be able to reproduce Skinner's textbook, *The Analysis of Behavior,* used in Skinner's Natural Science 114 class, in the Didak 501 format in time for the beginning of the semester.[19]

Delays continued with the machines as well as the programs. The launch of the Didak, initially scheduled for November, was pushed back to December. Skinner feared that Rheem was cutting back on a range of its business activities. (It had sold its semiconductor business to Raytheon the previous month.[20]) "Will the Califone Company and teaching machines be the next to go?" Skinner fretted. Rheem's vice president assured Skinner that Rheem had no plans to sell Califone or leave the teaching machine business."[21] But by the end of 1961, there was still no Didak for sale.

By the end of the first month of the new year, Rheem had reshuffled its management team, placing Gordon Mallatratt in charge of the teaching machine project. Skinner was delighted at first with this new leadership, but his enthusiasm quickly faded.[22] Mallatratt arranged for twenty Didaks to be shipped to Skinner, who—no surprise—found that they did not meet his expectations. "Many of the new Didaks are causing trouble," Skinner complained to Coons, "and if it were not for the fact that I have an excellent machinist here, we would be in serious difficulties in using them."[23] He then proceeded to list the problems: springs, brass bearings, counters, paper feeders—all of these were faulty in some way.

In February, Mallatratt again broached the subject of a revision to the original agreement Rheem had signed with Skinner three years earlier. "It is my understanding that your objective," he wrote to Skinner, "would be to increase your freedom to consult with other commercial clients in the field of teaching machines as well as to have the opportunity to arrange for the profitable commercialization of teaching machine concepts that you may develop in the future."[24] Mallatratt insisted that progress was being made at Rheem. A deal had been signed with Meredith and Basic Systems to obtain publishing rights to their programming materials, for example. He again invoked the figure of $300,000 that Rheem had invested in its teaching machine effort. "We are, of course, anxious to retain your services in connection with the further development of devices already in the prototype stage at Rheem Califone. However, this activity should not require an exclusive consulting arrangement and will permit you to realize your personal objective of increased freedom to work with any other commercial organization you may select." Another meeting was in order, Mallatratt suggested.

Skinner responded indignantly. He hadn't wanted to update or modify the original contract, nor did he have any interest in consulting with other commercial clients.[25] All he'd ever asked was for Rheem to carry out the terms of that initial agreement. Skinner scoffed at the $300,000 the company had spent—"It is certainly a high figure, considering what you have to show for it." He refused to take any blame for this expenditure, writing, "Much of Rheem's activity, for example, sending out machines for field testing prematurely, I have strongly advised against." Skinner then touted his own investment: "I am almost wholly responsible for current developments in the field of teaching machines, yet through my association I have been forced to remain almost inactive (in any effective way) for nearly three years. . . . Three years represent a much greater share of my working capital than $300,000 represent for Rheem." Skinner demanded precise details from Rheem on what any new or modified agreement with the company would entail: what would he be paid? Would royalties change? Could he build devices with other firms? Skinner informed Mallatratt he had no plans to travel to New York for a meeting with Rheem executives until he had more details that he could run by his attorney first.

Mallatratt replied a month later, apologizing for taking so long to respond. "It happens that we have been considering your letter in light of our appraisal of Rheem's overall position in the field of programmed instruction and did not want to respond hastily before we had reviewed our current thoughts."[26] The correspondence was carefully but brutally worded.

> I think you will agree that the whole field of programmed instruction has been changing very rapidly and will continue to change in the foreseeable future. While the original concept in this field was pioneered by you and continues to be the basis for all activities, the last two or three years has seen a major change in the relative importance of machine manufacturers compared with publishers. Recently

publishers of programmed texts have assumed a more dominant position in the industry. Whether this will continue or whether machines will achieve a better position remains to be seen.

Mallatratt reiterated Rheem's commitment to the development of the Didak 501, as well as to securing programming materials for use with the device, but he indicated that the company's patent attorneys were not confident that its proprietary rights could be extended to new teaching machines. "To be perfectly candid with you," he told Skinner, "we feel that devoting extensive resources to the research, development, and promotion of any machines other than the 501 could not currently be justified on the basis of sound business principles. We remain vitally interested in the other machines, but until the 501 has shown that teaching machines are the answer to the needs of at least some portion of the education market, we cannot justify full speed development and production of other machines."

The change in plans demanded an adjustment to the Rheem-Skinner contract, Mallatratt argued. And as if telling Skinner, after all this time, that there was no commercial market for the Didak 501 weren't enough, Mallatratt continued to twist the knife, stating, "In our view the present annual fee for your consultation services is no longer appropriate. If you accept this position, we would, of course, expect to surrender our exclusive rights to your consultation services in the area of teaching machines and programmed instruction and to free you to render such services to other interested parties. We would hope, however, to continue to have the advantage of your consultation as needed at an appropriate per diem rate." Mallatratt added that Rheem would like to adjust the royalty fees, eliminating the minimum fees it had once agreed to give Skinner.

Skinner responded rather promptly, writing back to Mallatratt just two days later. "We appear to differ on three points of considerable importance," he wrote furiously.[27] The first: "The diligence with which Rheem has carried out the terms of our agreement to date." The second: "The direction and scope of further activity in the field of teaching machines." Skinner's assessment about the role of publishers versus that of manufacturers vis-a-vis the future of programmed instruction was quite different from Rheem's. "It is true that publishers have stepped in quickly to promote programmed texts in spite of the extraordinary economic advantage offered by machines. They have been very clever about this and very deliberate, and I am sorry that I have never been able to convince Rheem of the importance of the issue. Moreover, as my current project makes clear, there are important areas in which instructional devices are absolutely essential." And finally, Skinner's third point: "The value of my services as a consultant in the teaching machine field." Skinner was the founder of the teaching machine movement, he insisted angrily. His services were invaluable.

"In view of these very substantial disagreements, it would appear we have reached a point at which we can be no further help to each other. I therefore suggest that we terminate our agreement as of April 1, 1962." Skinner returned the check that Rheem had cut for his consulting services for the second quarter of the year, and he asked that all his patents be reassigned to him. He indicated he would be willing to let Rheem continue to manufacture the 501, but he insisted that he would need appropriate royalties.

Mallatratt responded to Skinner a week later, with an attempt to negotiate with the professor.[28] He offered to reassign the teaching machine patents, and then Rheem would pay Skinner for an exclusive license to manufacture the Didak 501. Royalty fees would continue, although without the minimum quarterly guarantee. Rheem also sought exclusive rights to the commercial use of Skinner's name—both in conjunction with the 501 as well as with any other teaching machine under development. "We would hope that where you could give your full approval on the design of a machine we could indicate such approval in our advertising. In cases where the machine design does not meet with your approval but is based on your ideas and design we would use a phrase such as 'based on a design by Dr. B. F. Skinner' as you suggested." Mallatratt, as Skinner requested, also enclosed a check for $1,250 with the letter—Skinner's quarterly minimum royalty fee (a little over $10,000 in today's dollars).

Although Mallatratt's missive was meant to be conciliatory—at the very least, it made no mention of Skinner's statement about canceling the two parties' agreement effective April 1—Skinner was not impressed with Mallatratt's offer. "Since Rheem is unable to proceed with the development of teaching machines on the scale specified in our agreement, it will now be necessary for me to seek the help of another manufacturer in order to go forward on a scale and in a direction which seems to me advisable," he responded. "Any fragmentary arrangement with Rheem such as you suggest would be a severe handicap to me in working out any such relation." Skinner repeated his call to terminate the agreement, again, effective April 1, 1962—that is, eight days prior to the date of his letter.[29] He indicated that he'd be glad to give Rheem an exclusive license to manufacture the Didak 501, and he would be fine with foregoing the minimum royalty fee as long as there would be a way to terminate the license if Rheem stopped producing or selling the machine. "I assume that Rheem is not interested in manufacturing and distributing a machine unless it sells at least a thousand machines per year. Shall we agree that if in any six month period, my royalty falls below $2,500.00, the license will terminate?" Skinner clarified that there need not be any royalty plans for other devices because he had no plans of working with Rheem to develop any. "I also do not want to participate in any further consultation with Rheem—in part, because this would interfere with good working relations with another company—so that compensation need not be discussed." He continued, "I cannot agree to having my name used in connection with any other machine now under development at Rheem Califone. I agree to the use of the phrase 'based on a design by B. F. Skinner' in connection with the 501 with the stipulation that if I participate in the design of another machine operating on the same or similar principles for another company, my name could also be used by such a company in promoting such a machine." Skinner thanked Mallatratt for the check he'd sent but closed his letter by reiterating, "So far as I am concerned, our agreement has now terminated."

The two men corresponded throughout April, sending letters back and forth offering deals and conditions that were never acceptable to the other party. In May, Mallatratt gave Skinner two alternatives: continue his relationship with Rheem with an exclusive license or continue it with a nonexclusive one—the royalty payments would be determined accordingly.

Skinner drafted a short and terse response to the ultimatum. If those were his only choices, then frankly, he'd prefer to keep the terms of the original agreement in place. Skinner's attorney urged him to simply state that he was refusing the deal "for reasons I have stated in detail in earlier correspondence."[30]

Mallatratt remained hopeful that some sort of arrangement could be worked out. But there was no communication between Rheem and Skinner over the course of the summer of 1962.

In the fall, Skinner undertook a European speaking tour, which Rheem hoped would drum up publicity for the Didak 501 overseas. While there was some enthusiasm for the Harvard professor's theories, it appeared that, in Rheem's absence, many other manufacturers had already made headway in the European market. A. J. Budden, who worked in Rheem's London office, apologized for the "overwhelming publicity received by the AutoTutor," Norman Crowder's teaching machine. "This is due to the fact that U.S. Industries have an active company in England and have done a very good job of promotion," he said, trying to reassure Skinner that "psychologists and responsible people in the educational field in Britain tend to favour the Skinnerian techniques."[31] Despite the popularity of Crowder's machine abroad, when Skinner returned to the United States he informed Mallatratt that "it was I think a very successful trip. There is obviously a rapidly growing interest in teaching machines throughout Europe.

I saw several local products, and U.S. Industries' AutoTutor kept turning up. I was able to present the whole position in a conservative way which will take some of the steam out of the gadgeteers."[32]

Skinner also remarked to Mallatratt that he'd heard that the patent application for the Didak had been granted while he was away, and he'd read a report that Rheem was working on patenting another Skinner device, this one with an anti-cheating feature some educators had requested.

But when the new year rolled around, Skinner hadn't seen any paperwork confirming the Didak patent had been awarded, nor had he heard any more about the "no-cheating" patent either.[33] His frustrations again boiled over, and in February, Skinner wrote to Mallatratt with a carefully worded letter, drafted under advisement of his attorney.

> We both feel that the agreement should be terminated in accordance with its terms—namely ninety-days advance written notice. The licensing arrangement you suggest is not very attractive. As I have pointed out in the past, I must now begin working with another company interested in developing teaching machines on a broad scale. I cannot evaluate exactly the disadvantages which I would suffer from a continuing non-exclusive license to Rheem to manufacture a machine of my design. I am quite sure, however, that the terms you suggest would not be adequate compensation. If sales reported for the last quarter of 1962 are any indication of the future (and you have not indicated any intention to promote the machine more intensively), you are offering me less than one hundred dollars per quarter for a non-exclusive license. I also have no great confidence that Rheem will continue to market such a machine for any length of time so that the disadvantages I would suffer in working out other arrangements during the next few months would scarcely be compensated for at all. I am afraid I must ask you

to either continue the agreement in force or to terminate it completely.[34]

Rheem finally agreed. It was time to end the company's contract with Skinner—effective two years earlier on July 1, 1961.[35] On August 30, 1963, all patent rights were reassigned to Skinner.

Skinner later wrote that he regretted his earlier "frankness, my wilted-shirt attitude, my willingness to go on . . . a first-name basis. If I stood on my dignity, kept advice in reserve, acted the 'professor' according to a businessman's script, I would probably be consulted and listened to."[36]

Rivkin wrote to Skinner in mid-September that "I am glad that the Rheem matter at last appears to be resolved." The one piece of outstanding business: his attorney fees. Originally, Rivkin was to be paid 10 percent of receipts Skinner earned from the Rheem deal, he reminded his client. Under that agreement, which ended in October 1959, Skinner had paid Rivkin about three thousand dollars. But there was now no more revenue to share. "We find in reviewing the matter that between January 1, 1960 and August 31, 1963, we have spent the rather astonishing total of 107.9 hours on Skinner matters. Most of this was concerned by the intermittent attempts to renegotiate your agreement with Rheem. How do you think that this should be treated? Do you think that we should send you a statement for services rendered after October 15, 1959?"[37]

There would be no commercial success for B. F. Skinner in building his teaching machines. In the end, he was left instead with a very large attorney bill.

11

PROGRAMMED INSTRUCTION AND THE PRACTICE OF FREEDOM

At the close of World War II, B. F. Skinner had decided to write a utopian novel, inspired, as he later wrote in his autobiography, by a dinner conversation with a friend whose son and son-in-law were stationed in the South Pacific. "I began to talk about what young people would do when the war was over," he recalled. "What a shame, I said, that they would abandon their crusading spirit and come back only to fall into the old lockstep of American life—getting a job, marrying, renting an apartment, making a down payment on a car, having a child or two."[1] When asked what he thought they should do instead, Skinner said "they should experiment; they should explore new ways of living."

His novel was originally titled "The Sun Is But a Morning Star," a nod to the last line of Henry David Thoreau's *Walden*. Skinner believed that his book was written in the same spirit—"I think a good deal of the thinking is Thoreauvian, particularly the possibility of working out a way of life independent of political action. My attitude toward punishment and aversive techniques of control fits nicely with Thoreau's civil disobedience."[2]

The novel, published as *Walden Two*, followed "a standard utopian strategy," as Skinner characterized it.[3] "A group of people would visit a community and hear it described by a member."[4] That member, in this case, was Frazier, the founder of a commune called Walden Two, "a self-proclaimed genius who has deserted academic psychology for behavioral engineering, the new discipline upon which the community is based."[5] The novel described that community's structure—how it organized labor, property, family, and education—mostly through a dialogue between Frazier and Burris, the former's old university professor and a visitor to and skeptic of the community. It was, as Skinner's biographer Daniel Bjork puts it, "a bold extrapolation of the results of operant conditioning with animals, an imaginative effort to create a better way of life for humans."[6]

When *Walden Two* was published in 1948, interest was minimal. Indeed, Skinner had struggled to find a publisher, finally striking a deal with Macmillan, which accepted the manuscript only on the condition he would also write an introductory psychology textbook for the company. (That book, *Science and Human Behavior*, appeared in 1953.) Skinner was, after all, not a novelist. Although he'd always aspired to be a writer, he was a psychology professor and, in the mid-1940s, a relatively unknown one at that.

As with the entirety of Skinner's oeuvre, reactions to the novel were mixed—the book received some good reviews and some scathing ones. As he admitted in the introduction to a new edition in 1975, "one or two distinguished critics took the book seriously, but the public left it alone for a dozen years."[7] Sales of the novel were modest at first, although it did outsell Skinner's only other book at the time, his 1938 text *The Behavior of Organisms*.[8] But interest in *Walden Two* grew tremendously over the following decades. "Only 9,000 copies had been sold between 1948 and 1960," Skinner reported, "but in 1961 alone another 8,000 were sold; in 1962, 10,000; in 1963, 25,000; and in 1964, 40,000."[9] By the 1970s, annual sales had risen to 100,000.[10] Something in American culture had clearly shifted.

Of course, in the intervening decades, B. F. Skinner had become a household name and one of the best-known intellectuals in postwar America. As such, the public had become familiar with his ideas of behavioral engineering—certainly far more so than anyone would have been in 1948—in part through his work on teaching machines and programmed instruction and his steady stream of appearances on college campuses, in magazines, and on television. According to Skinner, however, there was "a better reason why more and more people began to read the book. The world was beginning to face problems of an entirely new order of magnitude—the exhaustion of resources, the pollution of the environment, overpopulation, and the possibility of a nuclear holocaust."[11] Although new biological and physical technologies could, perhaps, address some of these crises, nothing would really shift until human behavior was fundamentally altered. That would require *psychological* technologies, he contended. What Skinner had fictionalized in the mid-1940s "was no longer a figment of the imagination" by the late 1960s, he asserted.[12] The public, he believed, was clearly ready for his technologies of behavior.

While there might have been some affinity between Skinner's ideas in *Walden Two* and the Sixties counterculture, particularly among those interested in establishing alternative communities, it would be a mistake to see the youth movements of the era as readily supportive of the Harvard professor's technologies, particularly as they were applied to education.

Criticism of the education system had shifted in the 1960s—at least in certain circles—away from the likes of Admiral Rickover and James Conant and their calls to make the curriculum more intellectually rigorous for all students. The publication of A. S. Neill's *Summerhill: A Radical Approach to Child Rearing* in 1960 was particularly galvanizing—the book became an instant bestseller—for those who wanted schools to be less strict, less standardized.

The *Summerhill* book described a school of the same name, a private school Neill had founded in England some forty years earlier that practiced democratic education and a radical permissiveness with regard to its pupils' studies and socialization. Much of Neill's philosophy of education was explicit in its opposition to behavioral controls—not only those advocated by Skinner but also those that were a core part of most schools' traditional disciplinary practices. "When we consider a child's natural interest in things, we begin to realize the dangers of both reward and punishment," Neill wrote in *Summerhill*. "Rewards and punishment tend to pressure a child into interest. But true interest is the life force of the whole personality, and such interest is completely spontaneous."[13]

Both Neill and his school were relatively unknown in the United States, and yet the book sold 24,000 copies its first year. By 1970, it had sold over two million.[14] As the review in the *New York Times* heralded,

> Neill's book has appeared . . . under the following circumstances: our public-school system has failed; our most eloquent voices have been urging freedom and true communal values; our cognitive researches have indicated that compulsion, grading, testing, and regimentation are harmful; our parents are desperate, and our children bewildered. Obviously a man will be listened to who under these circumstances says, "Start new kinds of schools. Don't coerce the children. Don't test them and grade them. Don't pretend they are all alike. Have patience, and have faith in the innate powers of life. I have done it myself, and here's how."[15]

1960 also saw the publication of social critic Paul Goodman's book *Growing Up Absurd*, which examined the disaffection spreading through American society, particularly among young people. Goodman blamed this malaise on the organization of postwar society, dictated by corporations' push for "safe nonconformity and competitive individuality."[16] The school system served to foster despair among the youth, and as such was the target of much of Goodman's ire, not only in his writing but in his activism as well. (He was a founding member of the American Summerhill Society, started by the publisher of *Summerhill*, Harold Hart.)

Called out for specific criticism was a certain Harvard psychology professor who, according to Goodman, had recently devised a way to strip even more spirit and creativity from an already stultifying institution. "In the elementary schools," Goodman wrote,

> children are tested by yes-and-no and multiple-choice questions because these are convenient to tabulate; then there is complaint later that they do not know how to articulate

their thoughts. Now Dr. Skinner of Harvard has invented us a machine that does away with the creative relation of pupil, teacher, and developing subject matter. It feeds the child questions "at his own pace" to teach him to add, read, write, and "other factual tasks," so that the teacher can apply himself to teaching "the refinements of education, the social aspects of learning, the philosophy of it, and advanced thinking."

It was a nice idea, a fine promise, Goodman said, but he doubted that this mechanization would actually lead to a more meaningful education. Indeed, it seemed more likely to strip humanity from the classroom. "Who, then, will watch the puzzlement on a child's face and suddenly guess what it is that he *really* doesn't understand, that has apparently nothing to do with the present problem, nor even the present subject matter?" Goodman wondered. "And who will notice the light in his eyes and seize the opportunity to spread the glorious clarity over the whole range of knowledge; for instance, the nature of succession and series, or what grammar really is: the insightful moments that are worth years of ordinary teaching."[17] As one of the founders of Gestalt therapy, Goodman stressed the importance of social relations in the development of the self. How might the teaching machine actually serve, he wondered, to alienate and isolate the student?[18]

Goodman continued his criticism of Skinner and teaching machines in his 1964 book *Compulsory Mis-education*, a book that challenged what Goodman felt were the superstitions American held about education—that more of it was always better, for starters. Like many would-be reformers, Goodman invoked the language of national crisis—although for him this crisis was existential, not merely economic—in his call for change, chastising the "school-monks" who dominated policy and research circles under the guise of making things better but who never seemed to address the underlying inequality and oppression that plagued school life.

Increasingly, Goodman argued, compulsory education had come to mean not only mandatory attendance but the conditioning of students toward a particular set of bourgeois, bureaucratic values. "The public schools of America have indeed been a powerful, and beneficent, force for the democratizing of a great mixed population," Goodman admitted. "But we must be careful to keep reassessing them when, with changing conditions, they become a universal trap and democracy begins to look like regimentation."[19] Goodman noted that "terrible damage is done to children simply by the size and standardization of the big system"[20]—criticisms that, to a certain extent, echoed those of earlier reformers like Ben Wood who had also identified standardization as a dangerous erasure of individuality. Goodman pushed back on the kinds of reforms that had been proposed by James Conant—that is, making high schools larger and more centralized, with more "'enriched' curriculum."[21] Rather than compulsory schooling, Goodman wanted a voluntary, more self-directed education—a kind of education that he linked, through John Dewey, to a longer history of progressivism, one that was relevant and humanizing. "There is no right education except growing up into a worthwhile world."[22]

Echoing the pessimistic tone of *Growing Up Absurd*, Goodman argued that "the future prospects" for the country were gloomy.

> If the powers-that-be proceed as stupidly, timidly, and "politically" as they have been doing, there will be a bad breakdown

and the upsurge of a know-nothing fascism of the right. . . . The other prospect—which, to be frank, seems to me to be the goal of the school-monks themselves—is a progressive regimentation and brainwashing, on scientific principles, directly toward a fascism-of-the-center, 1984. Certainly this is not anybody's deliberate purpose; but given the maturing of automation, and the present dominance of the automating spirit in schooling, so that all of life becomes geared to the automatic system, that is where we will land.[23]

Automation, rather than liberating teachers and students from drudgery, was reshaping society to one of more finely tuned control, he cautioned.

Goodman recognized the pressures that had prompted schools to turn to mechanization: the need to educate the soaring number of students the system was now compelled to serve. "What happened to the schools during the tenfold increase [in attendance] from 1900 to 1960?" he asked. "Administratively, we saw, we simply aggrandized and bureaucratized the existing framework. The system now looks like the system then. But in the process of massification, it inevitably suffered a sea-change. Plant, teacher-selection, and methods were increasingly standardized."[24] Standardized and now, Goodman observed, subsumed into the bureaucracies of postwar capitalism.

For Goodman, as for many education reformers before him, the fix was "individualization." But this individualization needed to rest upon self-regulation and not a regulatory mechanism like the teaching machine. "In reconstructing the present system," he argued, "the right principles seem to me to be the following: To make it easier for youngsters to gravitate to what suits them, and to provide many points of quitting and return. To cut down the loss of student hours in parroting and forgetting, and the loss of teacher hours in talking to the deaf. To engage more directly in the work of society, and to have useful products to show instead of stacks of examination papers. To begin to decide what should be automated and what must not be automated, and to educate for a decent society in the foreseeable future."[25]

Goodman was not unilaterally opposed to automation. "My bias," he wrote, "is that we should maximize automation as quickly as possible, *where it is relevant*—taking care to cushion job dislocation and to provide adequate social insurance."[26] But, as he cautioned, "the spirit and method of automation, logistics, chain of command, and clerical work are entirely irrelevant to humane services, community service, communications, community culture, high culture, citizenly initiative, education, and recreation. . . . The dangers of the highly technological and automated future are obvious: We might become a brainwashed society of idle and frivolous consumers."[27] Goodman argued that by automating education through teaching machines, the country was on the path to do just that. Programmed instruction, despite its promises of individualization, still involved *programming* after all.

Goodman did wonder if there might be some appropriate, "psychotherapeutic" usage of programmed instruction, perhaps "for the remedial instruction of kids who have developed severe blocks to learning and are far behind." These students lacked confidence, and Goodman thought there might be some comfort "in being able to take small steps entirely at their own pace and entirely by their own control of the machine." Working alone on a teaching machine was isolating—it ran counter to Goodman's insistence that school be a community—but as long as the classroom remained a site of competition, stratification, and judgment, then it might be far less damaging for a student who'd fallen behind to "be allowed to withdraw from the group and recover. And this time can usefully and curatively be spent in learning the standard 'answers' that can put him in the game again."[28]

Meanwhile a related idea—that is, the use of programmed instruction for remediation—was being explored elsewhere by a different set of education activists. In Mississippi, Robert Parris Moses, a leader of the Student Nonviolent Coordinating Committee (SNCC), was keen to try programmed instruction as part of an adult literacy project he was planning in the South. That project was part of the civil rights campaign in Mississippi called "Freedom Summer," and according to Moses' biographer Eric Burner, it belonged "in the lineage of Highlander, the voter registration schools, and Nonviolent High. To learn and to come together for learning meant defiance, solidarity, self-awareness, and a capacity for further action."[29] But programmed instruction had a very different lineage—the psychology laboratories of colleges and universities—and the challenge, for Bob Moses, would be to reconcile these.

Moses had received his undergraduate degree at Hamilton College, a private college in New York State and one of the schools that had been actively experimenting with teaching machines in the late 1950s and early 1960s. (Indeed, B. F. Skinner had attended Hamilton as an undergraduate in the 1920s.) In 1963, Moses wrote to John Blyth, a former Hamilton professor who'd been involved in the development of the devices, about his ideas for a literacy project.

Blyth, who was at the time the director of the Programmed Learning Department for The Diebold Group, reached out in turn to several foundations, securing funding to develop programming materials for illiterate adults.[30] In late April, Moses returned to his alma mater, gave a talk to its Emerson Literary Society, and observed courses at the college that used programmed instruction. He hoped his visit would help to raise funds for SNCC, even though he admitted to Julian Bond that "the college campus is conservative."[31]

The funding for the literacy project was used, in part, to renovate a home in Greenville, Mississippi, which housed about a dozen volunteers who worked with nearby Tougaloo College on the literacy curriculum and project implementation. "Student and helper have a loose-leaf book," one organization described the Diebold program.

> On a page is a picture, for example, of a man and beneath, the word "man." The system assumes that the student already has a certain verbal ability and that he will recognize the picture. To make sure that the student responds to the written word as well as the picture, he is required to underline the world. The following is the answer page; on it, there are the picture and the word, and the word is underscored. With the development of the response to words, and to sounds, the system progresses to phrases—"the big man"—to simple sentences—"the man is big," with accompanying pictures.[32]

While the small, step-by-step process is recognizable as programmed instruction, it should be underscored here this was not a student working in isolation with a machine, but rather a student working with a tutor. That tutor needed to remain supportive and encouraging and "should not be critical," one SNCC staffer reminded volunteers, "particularly at the start. For many of the students, just being able to verbalize in this situation is progress that can easily be inhibited by a disapproving remark or facial expression."[33]

Despite the desperate need, the literacy program struggled to attract participants—not surprising, as adult education programs often do. The stigma is powerful. But in the Deep South in 1964, the risks were especially grave, and in June, John Blyth informed Tougaloo College officials that Diebold was pulling out of Mississippi. It was too difficult "keeping the project separate from Civil Rights activities," he told the college president.[34] Blyth wrote to one of the funders of the project, informing him of Diebold's decision: "The public image of the project as part of the civil rights movement has increased the difficulty in finding suitable test subjects. It is difficult at best to locate completely illiterate individuals who are willing to venture into the strange environment at Tougaloo. The difficulty is increased when they believe that they face the risks entailed in civil rights actions."[35]

The language of "test subjects" by Blyth stood in stark contrast to the view of the SNCC volunteers. They saw the adult education program as a crucial part of a larger effort not just to teach reading and writing, but also to organize and activate the Black community. Indeed, one of the primary goals of SNCC's literacy project was to help facilitate voter registration in Mississippi. In 1962, only 6.7 percent of the state's Black adults were eligible to vote because of requirements like poll taxes and literacy tests that were clearly aimed at preventing them from doing so.[36]

In the spring of 1963, while planning for the following summer's civil rights campaign, Moses had also met with Teaching Machines Inc.'s Ben Wyckoff at his home in Atlanta, Georgia, to solicit his advice on the project. According to

Len Holt in his chronicle of Freedom Summer, their discussion prompted Moses to think about the ways in which operant conditioning and positive behavioral reinforcement, and not just programmed instruction, could be leveraged in civil rights campaigns in the South. "Consciously and unconsciously, the idea must have fermented within Bob Moses," Holt recounted.

> It was related to the whole program of registration and political organizing in Mississippi. The most discouraging aspect of the work was the feeling that it would be so long before even the trickle of voters registered could participate and feel that the risks undertaken were worth it. Reinforcement. Because of the poll-tax law of Mississippi, it was not unusual for there to be a two-year delay between registration and the first date of being eligible to vote. To vote in state elections, one has to have paid a poll tax for two consecutive years. Poll-tax payments are received for only one year at a time and only during a special time of the year; no double payments are permitted. Hence, the reinforcement is poor. By the close of the summer of 1963, an idea had jelled which would improve the reinforcement: the Freedom Vote Campaign.[37]

The purpose of the Freedom Vote Campaign was to allow Black people to participate in a democratic process, even though they were prevented from participating in the official election. Every adult resident in the state was eligible to register and cast an alternate ballot in the Freedom Election. By Holt's account, some 93,000 Black adults, many voting for the first time, did so.[38] That unofficial election would eventually give rise to the Mississippi Freedom Democratic Party, which famously sought a seat at the Democratic National Convention in 1964, arguing that the "official" Mississippi Democratic Party had been illegally selected in a segregated voting process.

In addition to these voter registration efforts that were part of Freedom Summer, civil rights activists sponsored "Freedom Schools," a network of alternative education centers that offered the kind of teaching and learning that the public school system of Mississippi had refused to provide its Black population—an education that combined both intellectual and political development and one that expressly linked knowledge with power. According to historian Daniel Perlstein, there was some interest among the SNCC activists for not only new curriculum and new pedagogies, but also, via new technologies, "educational innovation" that included adopting programmed instruction for some of the courses.[39]

But in many ways, programmed instruction was antithetical to the work of the Freedom Schools, in which "the teacher's job was not simply to teach but rather to learn with the students."[40] The pedagogy of these schools was akin to that promoted by Brazilian educator Paulo Freire (who was, in 1964, imprisoned as a traitor by the military junta who'd staged a coup in his country): "problem-posing education," a dialogue with students and teachers in which knowledge is jointly constructed. This, Freire contrasted with the "banking model of education," "in which the scope of action allowed to the students extends only as far as receiving, filing, and storing the deposits."[41] Programmed instruction, for its part, seemed to have much more affinity with the latter. Indeed, as Paul Goodman cautioned, despite all the talk of teaching machines enabling the individualization of education, programmed instruction was more apt to strip away student agency and selfhood. "Programmed teaching adapted for machine use goes a further step than conforming students to the consensus which is a principal effect of schooling interlocked with the mass media," Goodman wrote. "In this pedagogic method, it is *only* the programmer—the administrative decision-maker—who is to do any 'thinking' at all; the students are systematically conditioned to follow the train of the *other's* thoughts."[42] He continued: "'Learning' means to give some final response that the programmer considers advantageous (to the students). There is no criterion of *knowing* it, of having learned it, of Gestalt-forming or simplification. That is, the student has no active self at all; his self, at least as a student, is a construct of the programmer."[43] "Even if behavioral analysis and programmed instruction were the adequate analysis of learning and method of teaching," Goodman argued, "it would still be questionable, for overriding political reasons, whether they are generally appropriate for the education of free citizens."[44]

The Freedom Schools were resolutely committed to this sort of education—to the education of free people. This meant that SNCC had to rethink not simply how it used programming materials, but how it conceived of the whole process and practice of programmed instruction. The power and agency, for its purposes, could not reside with the programmer or the teacher, particularly as neither were likely to be Black or even from the South. Moreover, the goal of the Freedom Schools was to transform society, not to be conditioned to conform to it, and not to be conditioned to conform to someone else's notion of transformation. To that end, SNCC's Mary Varela listed among the goals for the programming materials she developed for a literacy project in Alabama: "'to work with the identity problem by introducing Negro History' and 'to help an adult create a vision for himself as a political entity and as an agent for social change.'"⁴⁵ Varela also sought to expand her project so that community members could be trained to run it instead of SNCC staff.

According to Bob Moses, this local dialogue in the service of building the educational materials became more important than the methodology of programmed learning itself. "I had gotten hold of a text and was using it with some adults," he recollected, "and noticed that they couldn't handle it because the pictures weren't suited to what they knew. . . . That got me into thinking and developing something closer to what people were doing. What I was interested in was the idea of training SNCC workers to develop material with the people we were working with. . . . At that point it was not programmed learning; there was a great deal of interaction. What would have happened [if the materials had been reused] I'm not sure."⁴⁶

Despite Skinner's fantasies of a well-engineered and egalitarian society in his novel *Walden Two*, his prescriptive behavioral programming could never lead to freedom, activists discovered, as it sought to shape and control, denying agency to the people they sought to uplift. And agency was key to learning. "To be candid," Paul Goodman mused, "I think operant-conditioning is vastly overrated. It teaches us the not newsy proposition that if an animal is deprived of its natural environment and society, sensorily deprived, made mildly anxious, and restricted to the narrowest possible spontaneous motion, it will emotionally identify with its oppressor and respond—with low-grade grace, energy, and intelligence—in the only way allowed to is. The poor beast must do something, just to live on a little."⁴⁷ He added that "it is extremely dubious that by controlled conditioning one can teach organically meaningful behavior. Rather, the attempt to control *prevents* learning."[48] This attempt at control reduces students to "mere objects of scientific interest," Paulo Freire wrote in *Pedagogy of the Oppressed*, first published in Portuguese in 1968 and translated into English two years later.[49] "Scientific revolutionary humanism cannot, in the name of revolution, treat the oppressed as objects to be analyzed and (based on that analysis) presented with prescriptions for behavior," he insisted.

"There is a pathos in our technological advancement well exemplified by programmed instruction," Goodman concluded. "A large part of it consists in erroneously reducing the concept of animals and human beings in order to make them machine-operable."[50]

As the civil rights and youth movements of the 1960s progressed, there seemed to be a growing consensus among student-activists that this was what society was in fact conditioning them for: the machine. The most famous condemnation of this mechanical future came in December 1964 from University of California, Berkeley student Mario Savio, who'd spent his previous summer in Mississippi, registering people to vote and teaching at a Freedom School in McComb. When he returned to Berkeley in the fall, he'd discovered that the university had banned all political activity and fundraising. This curtailing of student free speech led to campus protests, and from the steps of Sproul Hall, Savio explicitly linked the functioning of the university with a mechanized dehumanization: "There comes a time when the operation of the machine becomes so odious, makes you so sick at heart, that you can't take part. You can't even passively take part. And you've got to put your bodies upon the gears and upon the wheels, upon the levers, upon all the apparatus, and you've got to make it stop. And you've got to indicate to the people who run it, to the people who own it, that unless you're free, the machine will be prevented from working at all."[51]

The machine was not merely a metaphor for an automated, exploitative society. It was a nod to the growing power of information technology, historian Steven Lubar argues.[52] And while the phrase "information technology" is more readily associated with computers, it would be wrong to exclude from this formulation the testing and teaching machines of the pre-computer era.

While Ben Wood had called, some thirty years earlier, for a "mechanical education" and had partnered with IBM to build the machines that could bring this about, the students of the 1960s did not experience the individualization that Wood had promised, but rather a bureaucratic uniformity, one they readily associated with the IBM punch card. (The cover of Raymond Callahan's 1962 book, *Education and the Cult of Efficiency*, features a punch card superimposed over a classroom.) University of California, Berkeley, used the punch card to, among other things, manage course registration, and the card became a symbol for students who felt that the university was a "knowledge factory," that they were "nothing but a cog going through preprogrammed motions."[53] The words printed on the punch cards—"Do not fold, spindle, or mutilate"—were seized upon by students demanding to be treated with more care than the paper cards that controlled the machines that sorted, assessed, and taught them.

12

AGAINST B. F. SKINNER

Skinner's autobiography, penned decades after the close of his business endeavors, oozes with resentment and contempt for the "rank commercialism" of the teaching machine industry, particularly for Rheem Manufacturing.[1] "I suffered from the treatment by businesses in the teaching machine era," he wrote, recalling how he would lay "awake nights writhing in anger."[2]

And yet the story of teaching machines almost always places the blame for the failure of the movement elsewhere. It was not the fault, for example, of slow-moving manufacturers who dragged their feet in getting the machines to market. It was not the fault of the publishers who opted to print programmed workbooks so that they needn't be tied to the proprietary design of a particular teaching machine. Psychologist Arthur I. Gates once told Skinner that the reason Pressey's Automatic Teacher was unsuccessful was neither Welch Manufacturing nor the Great Depression, but rather "the increasing popularity of several forms of 'progressive education' [that] was obviously hostile to the development of this type of alleged 'predetermined mechanical' learning."[3] Even Skinner would argue that it wasn't so much that Rheem or Harcourt Brace or IBM had hampered the teaching machine movement. Instead, he bemoaned that his behavioral technologies were "being kept out of our schools by false theories of learning [and] teacher unions who are Luddites and are afraid that this is going to deprive them of their jobs."[4]

Most often the decline in the popularity of teaching machines, and with this, Skinner's retreat from education technology's center stage, are attributed to two interrelated forces: cognitive science and the computer. Certainly, by the mid-1960s, academia, industry, and the press had turned their attention to the educational potential of computing technology—frequently echoing the claims made just a few years earlier about what teaching machines would do. "One can predict in a few more years," Stanford University professor Patrick Suppes wrote in 1966, "millions of schoolchildren will have access to what Philip Macedon's son Alexander enjoyed as a royal prerogative: the personal services of a tutor as well-informed and responsive as Aristotle."[5] The computer would be the tutor. The computer, as the teaching machine before it, would "individualize" education. Indeed, according to Suppes, "the computer makes the individualization of instruction easier because it can be programmed to follow each student's history of learning successes and failures and to use his past performance as a basis for selecting the new problems and new concepts to which he should be exposed next."[6] The computer was, in many ways, a reprise of Simon Ramo's push-button school.

By the time B. F. Skinner published *The Technology of Teaching* in 1968, a collection of new and old essays on programmed instruction and teaching machines, his enthusiasm for the "movement" had waned dramatically.[7] The revolution that he (and Pressey) had predicted had not come to pass. Schools had not adopted teaching machines; their practices had not been transformed by operant conditioning. "I think education is the greatest disappointment in my life," he opined.[8]

Perhaps it's no surprise then, *The Technology of Teaching*, Skinner's only book devoted to education, was not written for teachers or administrators. "I find myself writing for experimental analysts," he said. "I like to think that they will see the significance of [the teaching machine and programmed instruction]. . . . I am convinced that education as such cannot be changed."[9]

However, while he might have hoped *The Technology of Teaching* would appeal to researchers, Skinner had become increasingly alienated from many of his colleagues, particularly as cognitive science gained ground in psychology circles and behaviorism fell from favor. Although concerned, broadly speaking, with "the mind," some cognitive scientists had turned their attention to education specifically and to how children learned. Most notable in this shift perhaps was the publication of two books by fellow Harvard psychology professor Jerome Bruner: *The Process of Education* (1960) and *Toward a Theory of Education* (1966).[10] "Theories of learning . . . are destroying this country," Skinner told his biographer Daniel Bjork, a not-so-subtle jab at Bruner.

Even Sidney Pressey had soured on behaviorism. In a 1963 article in the *Journal of Applied Psychology*, Pressey charged that "the whole trend of American research and theory as regards [to] learning has been based on a false premise—that the important features of human learning are to be found in animals."[11] Drawing, as other cognitive scientists would do, on the work of Swiss psychologist Jean Piaget, Pressey challenged behaviorism for failing to adequately account for the developmental stages children pass through—and pass through without "so crude and rote process as the accretion of bit learnings stuck on by reinforcements."[12] "Far more remarkable than Skinner's pigeons playing ping pong," Pressey wrote, "is the average human scanning a newspaper—glancing about to find matter of interest to him, judging, generalizing, reconstruing, all in silent reading without overt respondings or reinforcings. Most remarkable of all is it to see learning theorists, hypnotized by the plausibilities of a neat theory, trying to teach that human as if he were a pigeon—confining his glance to the rigid slow serial peep show viewing of innumerable 'frames' each demanding that he respond and be reinforced."[13] As Pressey had long predicted, the teaching machine movement faced a crisis, not simply because of businesses' overpromising but because of behaviorism's.

Skinner had been quite critical initially of computers—unsurprising perhaps considering the run-ins he'd had with some of their manufacturers. But he did eventually change his mind. "When I go back and look at the machines I invented," he told an interviewer in 1984, "I can see that they were just efforts to do mechanically what can now be done much more smoothly with computers. Of course, computers can do a lot of other things too. But as for the whole notion of presenting material and evaluating an answer, computers can perform beautifully now."[14]

While he might have come to see the potential, Skinner was never fully convinced that computer-aided instruction would be designed "correctly." Indeed, the very phrase "computer-aided instruction" revealed a misunderstanding, he argued, about how best to use this new machinery. "The small computer is the ideal hardware for programmed instruction," he wrote in 1986.

> It is not functioning as a computer; it is teaching machine. It should be called a teaching machine. We have flying machines, sewing machines, and calculating machines—and a machine that teaches by arranging contingencies of reinforcement is a teaching machine. When computers were first used as teaching machines, their sponsors began to speak of "computer-aided instruction." That terminology is correct if teachers merely use computers to help them teach, but it is wrong when the computer does it all. We do not speak of computer-aided calculation. We use a calculating machine.[15]

Concerns about the performance of the American school system had been reignited, particularly with the publication in 1983 of *A Nation at Risk*, a report that charged that "the educational foundations of our society are presently being eroded by a rising tide of mediocrity that threatens our very future as a Nation and a people."[16] *A Nation at Risk* contended that the momentum for education reform following the launch of Sputnik had been squandered, and Skinner was quick to point out that he'd offered a solution to the problems of the education system decades before—a solution that had largely been ignored.

Skinner wrote a letter to the editor of *Science* in 1989 in response to an article on the poor math proficiency of US students, insisting that computers were unlikely to solve any sort of educational problem.[17] He related the story of his visit to the Roanoke classroom in 1960, how students were enraptured by their teaching machines and performed well on standardized tests. "Computers are now much better teaching machines," he acknowledged, but the problem, he argued, was that "the basic principles of programmed instruction are not yet being followed. Why not? Possibly because cognitive psychologists, claiming an esoteric knowledge of how students think opposed them. Students were to think as mathematicians thought. The result, of course, was the New Math. That was the age-old strategy underlying all changes in curricula: when students are not learning, teach them something else. The failure of the New Math is now clear."[18]

The "New Math" was arguably one of the most reviled curricular reforms of the post-Sputnik era. Although there was "no such stable or coherent thing as the new math," as Christopher Phillips argues in *New Math: A Political History*, the label was used to describe new pedagogies and textbooks that were designed to teach mathematical thinking in a new way—indeed to teach it as a form of thinking, not simply as a practice of memorization.[19] In some ways, the rise and fall of the New Math mirrored that of teaching machines—it was introduced in the late 1950s; its popularity peaked in the mid 1960s; it had been rejected by the 1970s. Along the way, the new curriculum became a symbol of misguided reform, one that continued to resonate with the public years later, as Skinner's letter, written in 1989, would suggest. Phillips contends that the demise of the New Math was more political than pedagogical, inseparable from the broader context—namely, the public attitudes toward educational institutions and educational expertise.

The "Back to Basics" movement that came in the wake of the New Math was quite amendable to the kind of lessons offered on teaching machines and, later, on computers: drills and memorization. But that does not fully explain Skinner's antipathy toward the New Math curriculum. This came, rather, from its ties to cognitive science, his academic rivals.

The origins of the New Math could be traced, in part, to a gathering in Woods Hole, Massachusetts, held in September 1959, as a response to the Soviets' launch of Sputnik, and to Harvard psychology professor Jerome Bruner, who directed the event, publishing its proceedings as *The Process of Education*. That book's most well-known claim: "any subject can be taught effectively in some intellectually honest form to any child at any stage of development."[20]

Skinner's rejection of cognitive psychology echoed his disdain, from much earlier in his career, of psychoanalysis. It reeked of "mentalism," he charged, as it relied on metaphors for understanding the brain—in the case of cognitive science, the metaphor of "information processing."[21] Behaviorism, on the other hand, accounted for "the whole organism" and "the world around it," Skinner argued.[22] It was not an abstraction. Cognitive science, Skinner sneered, was only "called a revolution because it is said to have overthrown behaviorism."[23] But it wasn't a revolt, he insisted. "It is a retreat."[24] Skinner contended that many of the insights of cognitive science could be reworded in behaviorist terms. He also argued that cognitive science was flawed and as such was partly responsible for the ongoing struggles of the US school system. "Several years before Sputnik," Skinner recalled in a 1985 article in the *British Journal of Psychology*, "in experiments with teaching machines and a system of programmed instruction based upon behavioural analysis of verbal behavior, it was shown that what was then taught in American classrooms could be taught in half the time and with half the effort. But classroom practices were not changed, largely because education remained (and, alas, still remains) committed to cognitive theories."[25]

In 1971, Skinner published his most controversial book, provocatively titled *Beyond Freedom and Dignity*. The book, which was in some ways a philosophical treatise on the ideas Skinner had examined in his novel *Walden Two*, argued that freedom was an illusion, a psychological "escape route" that convinced people their behaviors were not controlled or controllable.[26] "Autonomous man serves to explain only the things we are not yet able to explain in other ways," Skinner insisted. "His existence depends on our ignorance, and he naturally loses status as we come to know more about behavior."[27] The literature on freedom and dignity—in other words, much of Western philosophy—"stands in the way of future human achievement."[28] What was necessary to break from this false sense of autonomy, Skinner argued—here as throughout his work—was an analysis and, most importantly, a *technology* of behavior.

Published at the height of the counterculture movement and the Vietnam War, Skinner's book was an attack on the core tenets of democracy. Although Skinner had been a well-known public intellectual for decades, the press treated the behaviorist ideas expressed in *Beyond Freedom and Dignity* as shocking and new. Skinner appeared on the cover of *Time* in September 1971, white-haired and serious-faced, surrounded by four graphics that illustrated his most well-known inventions and ideas: a pigeon playing ping-pong, a rat pressing a lever in an operant conditioning chamber; the lush green pastoral commune of Walden Two; and a small hand pressing a button on a teaching machine. "B. F. Skinner says: We Can't Afford Freedom," read the magazine headline.[29]

The *Time* profile of the famous psychologist highlighted the number of critics that Skinner had attracted over the years, many of whom found behavioral reinforcement "philosophically distasteful and morally wrong."[30] Skinner had derided this "name-calling" in *Beyond Freedom and Dignity*, chastising his critics for their "fanatical opposition" to his work.[31] But the publication of the book simply served to fuel these objections. Not only was Skinner's behaviorism naïve, cruel, and intellectually bankrupt, claimed his critics, but also his arguments in *Beyond Freedom and Dignity* were totalitarian.[32] As Herbert C. Kelman, another Harvard psychologist, told *Time*, "For those of us who hold the enhancement of man's freedom of choice as a fundamental value, any manipulation of the behavior of others constitutes a violation of their humanity, regardless of the 'goodness' of the cause that this manipulation is designed to serve."[33]

The problem with Skinner's behaviorism wasn't just the science, it was also the politics. This was the crux of one of the most influential critiques of *Beyond Freedom and Dignity*: Noam Chomsky's review, "The Case Against B. F. Skinner," appeared in the *New York Review of Books* in 1971. It was not the first hostile review that the MIT linguist had written of Skinner's work; Chomsky lambasted *Verbal Behavior* in a

1959 book review.[34] "The Case Against B. F. Skinner" was also devastatingly brutal. "As to its social implications," Chomsky famously wrote, "Skinner's science of human behavior, being quite vacuous, is as congenial to the libertarian as to the fascist."[35]

In challenging Skinner's behaviorism, Chomsky insisted that science simply must investigate "internal states." By refusing to do so, "Skinner reveals his hostility not only to 'the nature of scientific inquiry' but even to common engineering practice"—to troubleshooting how machines function within, for example. "By objecting, a priori" to the examination of inner workings, "Skinner merely condemns his strange variety of 'behavioral science' to continued ineptitude."[36]

Even if it was inept, Chomsky contended, that didn't mean it was not dangerous. "There is nothing in Skinner's approach that is incompatible with a police state in which rigid laws are enforced by people who are themselves subject to them and the threat of dire punishment hangs over all." Skinner promised that behavioral engineering would "make the world safer," but Chomsky did not believe for a moment that a benign state, run by behavioral scientists, would be the result.

In many accounts, both behaviorism and Skinner found themselves largely discredited following the publication of *Beyond Freedom and Dignity*, thanks in no small part to Chomsky's book review. In psychology departments around the country, cognitive science had become the dominant approach.[37] But the academic debates were probably not the ones that came to shape the public's opinion of either behaviorism or Skinner. Most people would not have interpreted the failures of the New Math as a vindication of an early psychological approach. Most people would not have read Chomsky's book review. The public did, however, flock to watch a movie released (in New York City) in the closing days of 1971: Stanley Kubrick's *A Clockwork Orange*.[38]

To be fair, the film, based on Anthony Burgess's 1963 novel, did not depict operant conditioning. Skinner had always argued that positive behavioral reinforcement was far more effective than aversion therapy—than the fictional "Ludovico Technique" that *A Clockwork Orange* portrays.

In a futuristic Britain, Alex (played by Malcolm McDowell) is the leader of a gang of "droogs" who engage in a series of acts of "ultra-violence"—assault, rape, and eventually murder. Alex is caught and sentenced to prison for fourteen years. Two years into his sentence, he volunteers for an experimental treatment, proposed by the Minister of the Interior, which promises to rehabilitate criminals after just two weeks.

This treatment is the Ludovico Technique. Alex is strapped to a chair, his eyes are clamped open, he is injected with nausea-inducing drugs, and he is forced to watch violent and sexually explicit films while the music of his favorite composer, Ludwig van Beethoven, blares in the background. He is conditioned: the drugs, music, and the graphic depictions make him sick.

After two weeks of treatment, the Minister of the Interior demonstrates Alex's progress to a group of officials. Alex is provoked with physical violence and a naked woman; his only response is nausea. The Minister is triumphant, but the prison chaplain protests that the experiment has robbed Alex of his free will. "The boy has no real choice," he complains. "He ceases also to be a creature of moral choice." The Minister—his name, "Frederick," a nod to Skinner's middle name, "Frederic"—insists that Alex's mental processes are irrelevant. "We are not concerned with motive, with the higher ethics," he retorts. "We are concerned only with cutting down crime"—only, that is, concerned with *behavior*.[39]

The education columnist for the *New York Times*, Fred Hechinger—someone who had, a decade earlier, written favorably of Skinner's teaching machines—castigated both Kubrick and the film. "Any liberal with brains should hate 'Clockwork,' not as a matter of artistic criticism but for the trend the film represents. An alert liberal should recognize the voice of fascism."[40] Both Kubrick and McDowell responded furiously with letters to the newspaper, charging that Hechinger, not typically a film critic, had completely misconstrued the movie and its underlying ideas. The movie did not celebrate fascism, Kubrick asserted. It condemned the new "psychedelic fascism—the eye-popping, multimedia, quadrasonic, drug-oriented conditioning of human beings by other beings." "Mr. Hechinger is no doubt a well-educated man," Kubrick concluded, "but the tone of his piece strikes me as also that of a well-conditioned man who responds to what he expects to find, or has been told, or has read about, rather than to what he actually perceives 'A Clockwork Orange' to be. Maybe he should deposit his grab-bag of conditioned reflexes outside and go in to see it again. This time exercising a little choice."[41]

A decade after the publication of his novel and two years after the release of the film adaptation, Anthony

Burgess wrote a lengthy essay about *A Clockwork Orange*—his thoughts on crime and punishment and behavior modification, with particular attention to the connections in his novel to Skinner's *Beyond Freedom and Dignity*. "What I was trying to say," he wrote, "was that it was better to be bad of one's own free will than to be good through scientific brainwashing." Skinner wanted to demonstrate that the latter—or conditioning, at least—was necessary and could be benevolent. Burgess continued: "Our world is in a bad way, says Skinner, what with the problems of war, pollution of the environment, civil violence, the population explosion. Human behavior must change—that much, he says, is self-evident, and few would disagree—and in order to do this we need a technology of behavior."[42]

Skinner had called for a "technology of behavior" of "the right sort." "It is," Burgess admitted, "in the Skinnerian argument, conditioning of the wrong sort that turns the hero of 'A Clockwork Orange' into a vomiting paragon of non-aggression." But Burgess rejected that argument altogether. He did not believe there could be a right sort. He believed, he said, in people's freedom to make bad decisions. He believed in their rights and in their dignity and, thanks to his Catholic upbringing, in the possibility of their redemption.

Fascism in Europe, Burgess contended, had been "a kind of clockwork condition, a zestless ticking of the human machine." Skinner's machinery of behaviorism was poised to resurrect this condition. During the Nazi occupation of France, he argued, the people were at their "least free." But paradoxically, "they were at least free to recover a sense of the dignity of human freedom. There was the Resistance; there was the final and irreducible freedom to say no to evil.

This is a right not available in a society concerned with reinforcing behavior. That a man may be willing to suffer torture and death for the sake of a principle is a kind of mad perversity that makes little sense in the behaviorist's laboratory."

Skinner had said as much himself: his technology of behavior—and that included the teaching machine—was not interested in or committed to freedom.

CONCLUSION

"In spite of experimental evidence suggesting they were effective, schools failed to adopt teaching machines and programmed instruction in any large measure," Bill Ferster argues in his history of education technology.[1] It's a common pronouncement: while there was a flurry of interest in teaching machines in the 1960s, the devices were never taken up widely; as such, their influence was minimal. Despite all the promises of an "industrial revolution," of a more individualized, mechanical education, teaching machines disappeared. They changed nothing.

But is that right?

Historian of psychology Ludy Benjamin tells a slightly different story, observing the more cyclical nature of teaching machines' acclaim. "They emerged in the 1920s at the hand of Sidney Pressey and were largely confined to the dissertations of a handful of Ohio State University doctoral students," he writes. "They reappeared in the 1950s with the work of B. F. Skinner and enjoyed considerable popularity through the early 1960s. By the late 1960s, they had gone the way of hula hoops, only to be reincarnated in the personal computers of the 1980s."[2] Viewed this way, teaching machines were not a short-lived fad but rather a recurring trend. According to Benjamin, teaching machines simply ran, again and again, into that very same "cultural inertia" that Skinner and Pressey decried and blamed for the failures of their inventions.[3]

But it doesn't seem accurate to describe education in the twentieth century as inert. There were numerous reform efforts—some more amenable to technological intervention than others, no doubt. And resistance to teaching machines did not just come from schools, it came as well from the business sector and from the public at large.

Whatever the public's opinion on Skinner might have been by the mid-1970s, that did not mean that teaching machines were rejected because of their ties to the infamous behaviorist. Indeed, teaching machines and programmed instruction were never thoroughly repudiated. Nor were they supplanted by digital technologies—either by the first attempts at computer-assisted instruction in the 1960s or by the educational software of today. Rather, subsequent education technologies have continued to draw on many aspects of instructional design that teaching machines' inventors and advocates developed decades earlier—breaking lessons down into the smallest possible pieces of content, for example, giving students immediate feedback on their errors, and allowing them to move at their own pace until they've mastered a concept.

Some education technologists have recognized their pre-computer antecedents, although typically, when they give a nod to the past, it comes with assurances that their solutions surpass the capabilities of older theories or tools. Patrick Suppes, one of the names most closely associated with the computer-aided instruction of the 1960s and 1970s, once described himself as the "white knight of the behaviorists," for example, clearly implying he was poised to rescue programmed instruction from its past.[4] More recently, Dreambox Learning, an "adaptive learning" software company funded in part by Netflix CEO Reed Hastings, has described itself as "rooted in cognitive psychology, beginning with the work of behaviorist B.F. Skinner in the 1950s, and continuing through the artificial intelligence movement of the 1970s."[5] The history isn't quite correct or complete in Dreambox's telling, as this book has shown. But it is, at least, an acknowledgement that *something* came before. More often than not, as Sal Khan inadvertently reveals in his "History of Education" video, earlier pedagogies and technologies are utterly ignored—education has been "static to the present day"—as new developments try to position themselves as innovative and original.[6]

There were, no doubt, always critics of programmed instruction and teaching machines, even from within the field of education technology—and not merely because of the association with Skinner or behaviorism. Programmed instruction was often too dogmatic, charged the NEA's W. Lee Garner, and as such was unwilling to move in new directions or respond to new research, some of which did not seem to favor its core tenets.[7] As Paul Saettler writes in his sweeping history *The Evolution of American Educational Technology*, a lot of the "experimental evidence" for teaching machines just didn't hold up: "Many of the requirements originally based on theoretical grounds were not supported in practice. For example, the need for the student to make an overt response, the need for carefully sequenced frames, and the need for continuous and high rates of positive reinforcement were not demonstrated. Nor was it found necessary for all the students to go through the same set of frames in a linear sequence. But, even more devastating, students frequently found the materials boring."[8] Like much of education technology, research on the effectiveness of teaching machines was, at best, inconclusive.

Nevertheless, work on programmed instruction did extend well beyond the "end" of the teaching machine movement, although often this work occurred outside the school system. Many of its most prominent advocates continued their research in the field (or in new, adjacent fields). Thomas Gilbert, who'd built a teaching a machine for Bell Labs, and Lloyd Homme, who'd founded Teaching Machines Inc, went on to help establish the field of human performance technology, for example—a field that applied behavioral psychology to the workplace. Susan Meyer Markle, who'd worked with Skinner to develop the early IBM programmed materials, also became involved in that field, serving as the head of the International Society for Performance Improvement. Ben Wyckoff, another executive at TMI, wrote programs to help employees improve workplace communication. Allen Calvin, whose research on teaching machines was piloted by the Roanoke Public Schools, ran a series of education businesses, including one that offered "managed instruction" for struggling urban schools. Even Welch Manufacturing, the company that had failed to commercialize Sidney Pressey's teaching machine in the 1930s, stayed involved in education technology, and Richard Welch, the nephew of the company's founder, launched an organization called Learning Foundations to promote the automation of education. Clearly this list isn't exhaustive; but it does show that many of the key figures in the teaching machine movement did not suddenly stop working in teaching or training when the focus turned to computer-based education. Many of the ideas that propelled programmed instruction persisted and spread into new practices and new technologies.

As Joy Lisi Rankin demonstrates in her book *A People's History of Computing in the United States*, too often the history of computers focuses solely on the work of the tech industry and its engineers at the expense of the teachers and students who worked on the earliest computing systems. These were located at universities, after all. "The Silicon Valley mythology does us a disservice," Rankin argues. "It creates a digital America dependent on the work of a handful of male geniuses. . . . *It minimizes the role of primary and high schools, as well as colleges and universities, as sites of technological innovation during those decades* [emphasis mine]."[9] And by ignoring the latter in particular, the Silicon Valley mythology overlooks the importance of the history of education technology in establishing many of the conventions of computing.

Take PLATO (Programmed Logic for Automated Teaching Operations), for example. Often credited as the first computer-assisted instructional system, PLATO was developed at the University of Illinois Urbana-Champaign starting in 1960. But while it might have been the first *computer-*assisted teaching system, it was hardly the first *machine-*assisted one, and ideas of what such a system should look like were not made out of whole cloth. PLATO was described in early reports by its creators as a "teaching machine"—that is, the development of PLATO was not seen as a break from but rather a continuation of earlier work in automated education.[10] Although PLATO did eventually have other functionalities—it's sometimes called the first learning management system—its earliest lessons utilized well-established methods of programmed instruction, namely the intrinsic or branched programming of Norman Crowder. A student could "proceed at his own speed" through a sequence of slides, punching in a correct answer to move forward to the next question. When a student gave an incorrect answer, the machine would break "the original problem into a number of elementary steps. By answering each of these 'subquestions' the student is led step by step to the correct answer."[11]

If, as Rankin posits, the work of teachers and students on early computing systems helped form what the field would become, then it's clear that programmed instruction, rather than being roundly dismissed or replaced, is in fact constitutive of computing.

Just as the narratives about the history of computing tend to downplay the role of education technology in its development, these stories also seem to minimize the importance of behaviorism as a foundational theory, preferring to describe computing as "cognitive" instead. Arguably, the significance of behaviorism has become more evident in recent years as today's software has embraced "behavioral design," a phrase that perhaps obscures the connection to Skinner's earlier form of behavioral *engineering*. Stanford University psychologist B. J. Fogg and his Persuasive Technology Lab, for example, teach engineers and entrepreneurs how to build products—popular apps such as Instagram and Uber can trace their origins to the lab—that manipulate user behavior, cultivating a kind of conditioned response. "Contingencies of reinforcement," as Skinner would call them. "Technique," French philosopher Jacques Ellul would say. "Nudges," according to behavioral economist Richard Thaler, recipient of the 2017 Nobel Prize for economics.

These new technologies are purposefully engineered to demand our attention, to "hijack our minds," technology critic Nicholas Carr charges.[12] They're designed to elicit certain responses and to shape and alter our behaviors. Ostensibly all these nudges make us better people—that's the most positive spin to the story, at least, promoted in books like *Nudge* and *Thinking about Thinking*—much as Skinner tried to convince readers that his psycho-technologies would make the world a better place. In reality, many of these nudges are designed to get us to click on ads, to respond to notifications, to open apps, to stay on web pages, to scroll, to share—actions and "metrics" that Silicon Valley entrepreneurs and investors value.

There's a darker side still to this, Harvard Business School professor Shoshana Zuboff argues in her book *The Age of Surveillance Capitalism*, as this kind of behavior management has become embedded in our new information architecture. The Silicon Valley "nudge" is a market-oriented nudge. But as these technologies increasingly are part of media, scholarship, and schooling, it's a civics-oriented nudge too.

We have known for some time now that technology companies extract massive amounts of data from us in order to run (and ostensibly improve) their services. But increasingly, Zuboff contends, these companies are now using our data for much more than that: to shape and modify and predict our behavior—"'treatments' or 'data pellets' that select good behaviors," as one education technology executive describes it to Zuboff.[13] She calls this new use of data "behavioral surplus," a concept that is fundamental to her analysis of surveillance capitalism, which she claims is a new form of political, economic, and social power that has emerged from the "internet of everything."[14]

Google and Facebook are paradigmatic here, and Zuboff argues that the former was instrumental in discovering the value of behavioral surplus when it began, circa 2003, employing user data to fine-tune ad targeting and to make predictions about which ads users would click on. More clicks, of course, led to more revenue, and behavioral surplus became a new and dominant business model, at first for digital advertisers like Google and Facebook but shortly thereafter for all sorts of companies in all sorts of industries. And that includes education technology, of course—most obviously in predictive analytics software that promises to identify struggling students (such as Civitas Learning), in behavior management software that's aimed at fostering "a positive school culture" (like ClassDojo), and in adaptive learning software that promises—as the teaching machines did that came before it—to allow students to move through content at their own pace (Dreambox Learning, for example).

The subtitle of Zuboff's book, *The Fight for a Human Future at the New Frontier of Power*, underscores her argument—an argument that harkens back to the criticism of programmed instruction in the 1960s by Paul Goodman and others—that the acquiescence to these new digital technologies is detrimental to our future, to our freedom. These technologies foreclose rather than foster possibilities. And that certainly seems plausible, at least as Zuboff describes it—with our social media profiles scrutinized to adjudicate our immigration status, our fitness trackers monitored to determine our insurance rates, our reading and viewing habits manipulated by black-box algorithms, our devices listening in and prodding us as the world seems to totter toward totalitarianism.

Zuboff draws in part on the work of B. F. Skinner to make her case—his work on behavioral modification of pigeons, but also his larger theories about behavioral and social engineering, which she says are best articulated in *Walden Two* and *Beyond Freedom and Dignity*. By shaping our behaviors—through nudges and rewards—new, digital technologies increasingly circumscribe our ability to make decisions. They impede our "right to the future tense," Zuboff contends.[15]

Despite technology companies' growing influence in education, despite Zuboff's reliance on Skinner's behaviorist theories, and despite her insistence that surveillance capitalists are poised to dominate the future of work—not as a division of *labor* but as a division of *learning*—Zuboff has nothing much to say about how education technologies specifically might operate as a key lever in what she sees as a new form of social and political power.[16] (The quotation above from the "data pellet" fellow notwithstanding.)

Personalization, which Zuboff identifies as central to the predictive products of surveillance capitalism, comes from the collection of data "about your inner states, real-world context, and specific daily life activities . . . all in the service of successfully training the machines that they might better target market operations to each moment of life."[17] But Zuboff fails to acknowledge the elements of personalization-by-machine that did not originate in Google's headquarters. As this book has demonstrated, the goal of "personalization" is hardly new. Indeed, its origins even predate the kind of behavioral engineering that Skinner envisioned a technology of teaching could provide too. The work of Ben Wood, for example, underscores the efforts already underway in the early twentieth century to profile students through rigorous testing—personality, intelligence, aptitude, and subject matter testing—in order, in his words, to "individualize" education. And as Wood argued, this process necessitated a *machinery* of education. The machinery might be more modern, but the underlying desire to collect data and influence people is not.

Technologies of behaviorism are central to personalization. And as Silicon Valley has turned its attention to education reform, it has designed new teaching machines for "personalized learning" based on the kinds of extractive analytic practices Zuboff describes. Like programmed instruction before it, personalized learning promises that students can move at their own pace through lessons. However, with the enhanced data extraction and analytical capabilities of modern computing, today's new teaching machines now claim to know more about each student, claim to be able to respond more rapidly, more intelligently, more efficiently than a human teacher or even a human tutor could. What we find in personalization today is not merely an outgrowth of some new sociopolitical system called surveillance capitalism; rather personalization is the pinnacle of a long-running dream of education technology.

Teaching machines may then be one of the most important trends in the twentieth century—both in education and in technology—precisely because they were not a flash-in-the-pan, as some scholars have suggested, but a harbinger. Their ongoing influence can be found in the push for both personalized technologies and behavioral engineering. But teaching machines' most significant legacy might be, quite broadly, in the technocratic culture that they helped engender in education. That is, teaching machines were not merely aids to teaching. "These machines," Eugene Galanter argued in his report from the first teaching machine conference in 1958, "are a theory of teaching."[18]

If, as Raymond Callahan argued in his 1962 book *Education and the Cult of Efficiency*, the first half of the twentieth century saw reforms that demanded schools to be run like businesses, then the latter half saw efforts that viewed schools as "systems" that should be run like machines.

Decades before Simon Ramo fantasized about a push-button education and learning engineers, Ohio State University professor Werrett Wallace Charters, best known for his contributions to the idea of curricular development, asked, "Is there a field of educational engineering?" Charters's question, posed in 1945, signaled that a new approach to education was emerging that combined science—both the science of psychology and of management—with technological innovation.[19] This would eventually become the *cybernetic* approach to education, based on the idea that society could be engineered and "steered" through the science of human-machine interaction.[20]

Grounded in a theory of information that grew out of many scientists' wartime experiences, cybernetics offered a way to analyze—and, ideally, control—the behavior of humans and machines. Although often seen as the province of engineers, cybernetics had its adherents in education circles, particularly as observers felt schools were struggling to adapt to "the technological age."[21]

Much as World War I helped shape the practices of early education psychologists and hastened the spread of intelligence testing, World War II had oriented the field of psychology toward weapons—or at least toward systems of war. Harvard psychology professor Jerome Bruner, for example, served in the Psychological Warfare Division, and when he gathered his scientist colleagues for the Woods Hole conference in 1959—a conference sponsored in part by the US Air Force—the language of education reform was explicitly the language of weapons systems. One report, commissioned by the National Academy of Sciences and shared at the event argued, for example, that "The goals of education . . . expressed in terms of the human functions and tasks to be performed can be exactly and objectively specified as can the human functions and tasks in the Atlas Weapons Systems."[22] The model for change in education was technical. "In our present day society, tremendous forward strides have been made in the design and development of new technical integrations of men and machines in the form of system," the Apparatus of Teaching group reported to Woods Hole conference attendees. Although education was not quite the same as "the most spectacular examples of system design . . . found in complex military situations," it was nonetheless "the kind of complex organic enterprise the improvement of which can be aptly planned according to system development principles."[23] By engineering the functions of teachers and machines, then, the entire system of education could be optimized.

This technocracy that emerged post-Sputnik meant a political shift in decision-making power in education (and elsewhere). Major funders of the push for education technology—the Ford Foundation in particular—largely bypassed teachers in their efforts—"a sign of this policy to short-circuit the profession," education professor James D. Finn observed in 1960.[24] For all the talk from B. F. Skinner, Sidney Pressey, and others that teaching machines were never meant to replace teachers, Finn recognized that new education technologies were likely "another instrument of Neo-Technocracy" that "forecast even more loss of control by the existing pre-technological profession."[25] "Expertise" shifted elsewhere. As Charters and Ramo had predicted, it shifted to the engineer, to the technologist.[26]

The reshaping of education into a technocracy wasn't simply about reorganizing labor and expertise. It also meant a change in knowledge—in how thinking and knowing (and by extension, teaching and learning) were conceived. Knowledge is a system. Thinking is a machine.

For James Finn, for example, education technology was not merely a collection of devices—the teaching machine or the film strip or the radio, for example—but rather, "it is a process and a way of thinking."[27] Machines "must be thought of in connection with systems organizational patterns, utilitarian practices and so on."[28] The new technocratic culture meant that education would be automated—literally or metaphorically, it did not matter. School had become a "system" to be engineered and controlled.

In the closing essay to the 1964 collection *Programs, Teachers, and Machines*, Northwestern University education professor Daniel Tanner invoked cybernetician Norbert Wiener to caution against this obsession with automation and this recasting of education as a mechanical endeavor. "Many critics of education are impatient with the inefficiencies of our schools," Tanner admitted. "They decry the tendency for education to lag far behind industry in automation. They want to see educational expenditures reduced dramatically through the use of highly efficient autoinstructional devices. But we must bear in mind that while the product of industry is an automobile, a refrigerator, or a washing machine, the product of education is a human being."[29]

Viewing the student as a product—and not, say, as the subject—of education reflects the long-running belief that schools were factories. But even with newer approaches to theorizing education—"systems thinking"—students were still seen as objects to be controlled, their behaviors to be shaped. Tanner, like some of the activists of the 1960s, resisted this framework and he tried to warn fellow education technologists about the uncritical adoption of teaching machines—both the devices and the ideology around them:

> While autoinstructional technology may prove invaluable for improving the efficiency of factual and skill-type learning, we must appreciate the limitations as well as the

potentialities of these devices. Although it is argued that teaching machines provide for individualized instruction by permitting each student to progress at his own rate of speed, programed learning actually represents a mass standardization of content and process in education. The teaching machine requires absolute uniformity of interpretation and response on the part of the learner. Even the textbook does not require this. The student reading a text must identify and sort out relevant material for himself, while, on the other hand, the teaching machine does all this for the student. The learning process should not be made as difficult as possible for the student, but we need to keep in mind that the teaching machine atomizes and predigests a great deal of the instructional material. Relatively little latitude is left for individual interpretation and analysis in the process of "operant conditioning." The learner is not permitted to develop a style of inquiry of his own. He must simply confirm to the style of the programmer. Under "operant conditioning" the student is not in control of the programed material. Instead, he is under the control of the program.[30]

The MIT mathematician Seymour Papert would echo this sentiment a few years later in his 1980 book *Mindstorms*, rejecting the intellectual compliance that computer-assisted instruction demanded from students. He observed that "in most contemporary educational situations where children come into contact with computers the computer is used to put children through their paces, to provide exercises of an appropriate level of difficulty, to provide feedback, and to dispense information. The computer programming the child."[31] Papert had a different vision of learning—constructionism—in which a child would construct knowledge rather than receive knowledge. In the late 1960s, Papert, along with colleagues Cynthia Solomon and Wally Feurzig, had developed the programming language LOGO, which was aimed at introducing children to computational logic. Children could use LOGO to "teach" a robotic turtle with a set of instructions—forward 10, right 90—and the turtle would echo these as a series of turns and pen marks on the floor, as such expressing geometry physically and enabling an embodied type of reasoning, Papert argued. "The child programs the computer."

Computers were powerful objects to think with, Papert contended. "In teaching the computer how to think, children embark on an exploration about how they themselves think."[32] But how the LOGO turtle expressed its "thinking" was arguably as much a behavioral act as it was a reflection of any sort of cognitive progress. And if this is the type thinking—"computational thinking"—that children are supposed to see as a model for their own, then it appears that epistemology is recast. It's not simply that the educational system is a machine in this technocratic vision; the mind is one too.

Students are taught, as Theodore Roszak cautioned in his 1986 book *The Cult of Information*, to think like computers. As this type of thinking becomes the privileged way of knowing and thinking and moving through the world, teaching and learning become the purview of the machine. "In contrast," Roszak wondered, "who will teach them to think any other way?"[33] Thus, the phrase "teaching machines" takes on new meaning: this is the work of computer scientists who "teach" machines, those who specialize in "machine learning." And as machines are purported to "think" and to learn, our minds now too are imagined as machines, and our educational endeavors are conceived as systems to engineer.

This was the fear of the French philosopher Jacques Ellul, who wrote in his 1954 book *La technique ou l'enjeu du siècle* ("Technique or the stake of the century," published in English as *The Technological Society* a decade later) a devastating critique of the ways in which society had become utterly subsumed by "technique." By this, he meant more than simply "technology" or machines; "technique," he argued, was "the totality of methods rationally arrived at and having absolute efficiency in every field of human activity."[34] Technique had become "the consciousness of the mechanized world."[35] Society, business, politics, education—all these, all institutions and practices, have become transformed by technique, which has steered them all toward efficiency, rationality, numeracy, artificiality, profit.

Ellul identified several areas in which schooling, often under the guise of humanism, consisted of "a profound and detailed surveillance of the child's activities, a complete shaping of his spiritual life, and a precise regulation of his time with a stop watch; in short in habituating him to a joyful serfdom."[36] Students were being shaped and conformed to society's demands—not just in terms of their intellect, but in terms of their very personality—and those demands were increasingly technical. Indeed, the importance of intellectualism was fading, Ellul contended. The twentieth century wanted technicians, not critics. As such, "the human brain must be made to conform to the much more advanced brain of the machine. And education will no longer be an unpredictable and exciting adventure in human enlightenment, but an exercise in conformity and an apprenticeship to whatever gadgetry is useful in a technical world."[37] Education will be an apprenticeship to the thinking machine, to machine learning, and to the teaching machine.

There is a danger, however, in believing this mechanization is inescapable, that teaching machines—whether wooden or plastic or digital—are inevitable, just as there is folly in believing that the history of education proceeds on a straight line from the Prussians to Khan Academy. "The myth of technological and political and social inevitability is a powerful tranquilizer of the conscience," MIT professor Joseph Weizenbaum cautioned in 1976 in his critique of computer science titled *Computer Power and Human Reason*. "Its service is to remove responsibility from the shoulders of everyone who truly believes in it."[38]

"There *are* actors," Weizenbaum insisted, not merely systems or machines that operate without our understanding or control. Moreover, these actors—the ones who shape the practices and institutions of education—are not only the entrepreneurs, engineers, and reformers who hope to transform it into a more automated system. Education is a civic responsibility. And even when certain actors seem powerful in their desire to build their machinery of education, they are, as this book has hopefully shown, as likely to bumble their visions as capitalize on them.

There has always been resistance to teaching machines and to the technocracy in which they are embedded. Sidney Pressey experienced it; B. F. Skinner experienced it too. And perhaps it's worth repeating that that resistance did not come only from disgruntled educators. There were skeptics within and without educational institutions. If we reject teaching machines and technologies of behavioral control in education, we certainly won't be the first to do so.

From the history of refusal, we can see when students and teachers and communities protested attempts to engineer them, into either enlightenment or submission. From the alternatives they imagined and built—most notably, perhaps, the Freedom Schools, we can glean ways to construct and share knowledge that depend on humans rather than machines, liberating us from the efficient control of the "Skinner box." These practices privilege the much messier forms of teaching and learning, forms that are necessarily grounded in freedom and dignity.

ACKNOWLEDGMENTS

It's been almost a decade now since "The Year of the MOOC." In case you have forgotten or are lucky enough to never have known, the acronym means "massive open online course," and this was a trend that, according to some pundits at least, was poised to end higher education as we know it.

It's been almost a decade, too, since I started thinking about writing a book on teaching machines. And I will be the first to admit that the book's focus has probably shifted and changed and improved far more in the intervening years than the MOOC providers' pedagogy or their technology.

In December of 2012, I was invited to Palo Alto for a small gathering to discuss the future of teaching, learning, and technology. The event was organized by Sebastian Thrun, the founder of the MOOC startup Udacity. Or, at least, Udacity paid for the attendees' travel and lodging and food. At the end of the weekend, Thrun took us for a spin in the Google self-driving car, his research specialty before he'd turned his focus to online education. As the car steered itself along I-280 and I listened to Thrun explain the data collection and analysis that had gone into getting the car to navigate the road on its own, I realized that there was a great deal of similarity between how Thrun imagined teaching students and how he imagined teaching machines. (That double meaning of "teaching machines" seemed significant.) In Thrun's explanation of the workings of the autonomous vehicle was the pure distillation of Silicon Valley's mindset toward education technology. It struck me as the perfect example of how so many entrepreneurs and engineers think about schooling and the role of technology in reshaping it: that is, if you collect enough data—lots and lots and lots of data—you can automate teaching and learning. And the world will be better for your engineering of the system.

This book sat on the proverbial back burner for a very long time—until I was awarded a Spencer Education Journalism Fellowship and spent the 2017–2018 school year at Columbia Journalism School. There, I took Sam Freedman's famous book-writing class, and there, this project finally started to take shape. Thank you, first and foremost, to the Spencer Foundation for its support for education journalists, and to the faculty at the Columbia Journalism School and the Teachers College who made my year there so memorable. A special thank you to Peg Tyre and Greg Toppo, who encouraged me to apply for the fellowship—something that, as a freelance writer without an institutional affiliation, I didn't think would be attainable. Thanks to Sam Freedman and everyone in that amazing class. (So many great books will come out of that seminar.)

This book draws heavily on letters and memoranda donated to university and organizational libraries, and my research would not have been possible without the help of archivists. Thank you to the archivists at the Ohio State University Archives, Harvard University Archives, Stanford University Archives, and Educational Testing Service Archives, who wheeled out box after box of materials on Sidney Pressey, B. F. Skinner, Patrick Suppes, and Ben D. Wood, respectively.

The Spencer Fellowship gave me access to the Columbia University library—if you've ever tried to conduct research without access to a research library, you know how frustrating it is that so much scholarship is locked up behind a paywall. Thanks to everyone who emailed me PDFs of articles I couldn't find online. And thank you in particular to Amy Collier and Noraya Razzaque, who helped me locate many non-digitized and hard-to-find materials.

I am deeply grateful for Susan Buckley at the MIT Press, who believed in me and believed in this project when it was still a very rough outline, and for everyone at the publisher, particularly Virginia Crossman and Julia Collins for their editing finesse. Thank you to Justin Reich and Mike Caulfield, who read early versions of the manuscript and helped me shape the story and the analysis. Some of these ideas here were first test-driven in talks I delivered at Austin Peay State University and the University of California, Santa Barbara. Thank you to the audiences who humored me when my thoughts on teaching machines were still fairly jumbled. My apologies for all the places in this book where they still are.

Thank you to my teacher-friends who were willing to fact-check certain passages pertaining to their areas of expertise—thanks, especially, to Frank Noschese for help with the physics of pencil marks. Thank you to my writer-friends who offered me kind encouragement when things seemed bleak, particularly Jesse Stommel, Sean Michael Morris, Kate Bowles, and José Vilson. Thank you to my reader-friends who reminded me how excited they were to read this book. I know you've been waiting a while, and I hope I did not disappoint you.

Most of all, thanks to Kin Lane, who has been here for this whole ride (including the ride in that Google self-driving car). He has been the best travel partner I could ever imagine, moving with me to New York City for my fellowship and moving back with me to the West Coast when our Upper West Side apartment sprung a disastrous leak. Kin has been the best thinking partner, too. He has listened to me talk through all aspects of this project and has supported me through the darkest days of book-writing. I love you, Kin.

To my dearest Isaiah: how I wish you were here to celebrate this publication with me. I will love and miss you forever.

NOTES

INTRODUCTION

1. Salman Khan, "Let's Use Video to Reinvent Education," March 2011, produced by TED, 20:20, https://www.ted.com/talks/sal_khan_let_s_use_video_to_reinvent_education.

2. Brian Urstadt, "Salman Khan: The Messiah of Math," *Bloomberg Businessweek*, May 19, 2011, https://www.bloomberg.com/news/articles/2011-05-19/salman-khan-the-messiah-of-math; Annie Murphy Paul, "Salman Khan: The New Andrew Carnegie?," *Time*, November 16, 2011, http://ideas.time.com/2011/11/16/salman-kahn-the-new-andrew-carnegie/; Will Oremus, "Salman Khan, Founder of Khan Academy," *Slate*, August 2, 2011, http://www.slate.com/articles/technology/top_right/2011/08/salman_khan_founder_of_khan_academy.html; Clive Thompson, "How Khan Academy Is Changing the Rules of Education," *Wired*, July 15, 2011, https://www.wired.com/2011/07/ff_khan/; Gregory Ferenstein, "How Bill Gates' Favorite Teacher Wants to Disrupt Education," *Fast Company*, February 17, 2011, https://www.fastcompany.com/1728471/how-bill-gates-favorite-teacher-wants-disrupt-education.

3. Salman Khan, *The One World Schoolhouse* (New York: Twelve Books, 2012), 78.

4. Salman Khan and Michael Noer, "The History of Education," November 1, 2012, produced by *Forbes*, YouTube 11:27, https://youtu.be/LqTwDDTjb6g. Excerpts from this video used with permission from *Forbes*.

5. Richard Barbrook and Andy Cameron, "The Californian Ideology," *Mute* 1, no. 3 (September 1, 1995), http://www.metamute.org/editorial/articles/californian-ideology.

6. Charles Duhigg, "Did Uber Steal Google's Intellectual Property?," *New Yorker*, October 22, 2018, https://www.newyorker.com/magazine/2018/10/22/did-uber-steal-googles-intellectual-property.

7. David Tyack, "The Future of the Past: What Do We Need to Know about the History of Teaching?," in *American Teachers: A Profession at Work*, ed. Donald Warren (New York: MacMillan, 1989), 408.

8. Sidney L. Pressey, "A Simple Apparatus Which Gives Tests and Scores—and Teaches," *School and Society* 23, no. 586 (March 1926): 373.

9. Jacques Ellul, *The Technological Society* (New York: Alfred A. Knopf, Inc, 1964), 349.

10. Francis Fukuyama, "The End of History," in *The National Interest* 16 (1989): 3–18.

11. L. Paul Saettler, "The Origin and Development of Audio-Visual Communication in Education in the United States" (PhD diss., University of Southern California, 1953); L. Paul Saettler, *A History of Instructional Technology* (New York: McGraw-Hill, 1968).

12. L. Paul Saettler, *The Evolution of American Educational Technology* (New York: McGraw-Hill, 1990).

13. Saettler, *A History of Instructional Technology*, 267.

14. Kevin Kelly, *What Technology Wants* (New York: Viking Press, 2010).

15. Larry Cuban, *Oversold and Underused: Computers in the Classroom* (Cambridge, MA: Harvard University Press, 2003).

16. John Blyth, "Teaching Machines and Human Beings," in *Teaching Machines and Programmed Learning: A Source Book*, ed. A. A. Lumsdaine and Robert Glaser (Washington, DC: National Education Association, 1960), 408.

17. C. J. Kirsch, "The Germomat III System: Made for the Factory and the Laboratory," (forthcoming).

CHAPTER 1

1. B. F. Skinner, *A Matter of Consequences* (New York: Alfred A. Knopf, 1983), 64.

2. Skinner, *A Matter of Consequences*, 65.

3. Charles Ferster and B. F. Skinner, *Schedules of Reinforcement* (New York: Appleton-Century-Crofts, 1958).

4. B. F. Skinner, *Particulars of My Life* (New York: Alfred A. Knopf, Inc., 1976), 301.

5. B. F. Skinner, *The Shaping of a Behaviorist* (New York: New York University Press, 1979), 4.

6. John B. Watson, "Jung as Psychologist," *New Republic* (November 7, 1923), 88.

7. B. F. Skinner, *About Behaviorism* (New York: Vintage Books, 1976), 18.

8. Skinner, *About Behaviorism*, 18.

9. Alexandra Rutherford, *Beyond the Box: B. F. Skinner's Technology of Behavior from Laboratory to Life, 1950s–1970s* (Toronto: University of Toronto Press, 2009), 162.

10. Rae Goodell, *The Visible Scientists* (Boston: Little, Brown and Company, 1977).

11. Rutherford, *Beyond the Box*, 23, 153.

12. Rutherford, 153.

13. B. F. Skinner, "The Science of Learning and the Art of Teaching," *Harvard Educational Review* 24, no. 2 (1954): 86–97.

14. B. F. Skinner, "The Science of Learning and the Art of Teaching," in *The Technology of Teaching* (Acton, MA: Copley Publishing Group. 2003), 9.

15. Skinner, *The Shaping of a Behaviorist*, 262.

16. Skinner, 274.

17. Skinner, 271, 274.

18. Skinner, 279.

19. Skinner, 274.

20. "Pigeons Play Piano and Do Other Smart Things," *Worcester (MA) Gazette,* June 14, 1950.

21. Skinner, "The Science of Learning and the Art of Teaching," 14.

22. Skinner, 15.

23. Skinner, 17; Skinner, *A Matter of Consequences*, 69.

24. Skinner, "The Science of Learning and the Art of Teaching," 22.

25. Skinner, *A Matter of Consequences*, 69.

26. Skinner, "The Science of Learning and the Art of Teaching," 6–27.

27. "Teaching by Machine," *Science News Letter*, July 17, 1954, 38.

28. Earl Ubell, "Machine Teaches Kids Arithmetic Painlessly," *New York Herald Tribune,* June 29, 1954; "Coming Up: Machine Age in Teaching," *Worcester (MA) Telegram*, June 29, 1954; Associated Press, "The Atomic Age Hits Teachers," *World Telegram*, June 29, 1954; "Miracle Gadget Makes Boys Like Arithmetic," *Boston Herald,* June 29, 1954.

29. Edwin G. Boring, "CP SPEAKS," *Contemporary Psychology* 2, no. 12 (December 1957): 312–313.

30. Ludy Benjamin, "A History of Teaching Machines," *American Psychologist* 43, no. 9 (September 1988): 703–712.

31. Horace B. English to Edwin G. Boring, January 30, 1958, Sidney Leavitt Pressey Papers, University Archives, Ohio State University Libraries (hereafter, Pressey Papers).

32. Edwin G. Boring to Horace B. English, February 1, 1958, Pressey Papers.

33. Skinner, *The Shaping of a Behaviorist*, 73.

34. Edwin G. Boring to Horace B. English and Sidney L. Pressey, February 3, 1958, Pressey Papers.

35. Edwin G. Boring, "CP SPEAKS," *Contemporary Psychology* 3, no. 6. (June 1958): 152.

36. Boring to English and Pressey, February 3, 1958, Pressey Papers.

37. B. F. Skinner, "Teaching Machines," *Science* 128, no. 3330 (October 24, 1958): 969.

38. Sidney Pressey to B. F. Skinner, July 30, 1954, Pressey Papers.

39. Sidney Pressey to B. F. Skinner, September 22, 1954, Pressey Papers.

40. Sidney Pressey to B. F. Skinner, February 23, 1955, Pressey Papers.

41. Skinner, *A Matter of Consequences*, 71.

CHAPTER 2

1. "Welfare of the World Depends on Science, Coolidge Declares," *Washington Post*, January 1, 1925, 1.

2. George Arps, "Report of the College of Education for the Year Ending June 30, 1926," *Fifty-Sixth Annual Report of the Board of Trustees of The Ohio State University for the Year Ending June 30, 1926* (Columbus: The Ohio State University Press, 1926), 58.

3. Sidney Pressey, "A Simple Apparatus Which Gives Tests and Scores—and Teaches," *School and Society* 23, no. 586 (March 1926): 373.

4. Local Board for Division 31, State of Massachusetts, "Certificate of Discharge Because Physically Deficient," August 6, 1917, Sidney Leavitt Pressey Papers, University Archives, Ohio State University Libraries (hereafter, Pressey Papers).

5. Raymond Callahan, *Education and the Cult of Efficiency* (Chicago: University of Chicago Press, 1962), 105.

6. Victoria Luther, "Circling the World with Psychological Supplies," *Industrial Psychology Monthly* 2 (1927): 12–16.

7. Stephen Petrina, "The 'Never-to-Be-Forgotten Investigation': Luella W. Cole, Sidney L. Pressey, and Mental Surveying in Indiana, 1917–1921," in *History of Psychology* 4, no. 63 (2001): 246.

8. Sidney Pressey and Luella Cole, *Introduction to the Use of Standard Tests* (Yonkers, NY: World Book Company, 1922).

9. Sidney Pressey, "Sidney Leavitt Pressey," in *A History of Psychology in Autobiography*, ed. E. G. Boring and G. Lindzey (East Norwalk: Appleton-Century Crofts. 1967).

10. Sidney Pressey, "A Simple Self-Recording Double-Action Multiple Choice Apparatus," *Psychological Bulletin* 22 (1925): 111.

11. "Machine Tests Intellect," *Wapakoneta Republican*, February 5, 1925.

12. Sidney Pressey to C. M. Stoelting Co., May 1, 1926, Pressey Papers.

13. Sidney Pressey to the A. B. Dick Company, December 21, 1925, Pressey Papers.

14. Sidney Pressey to the Remington Typewriter Company, April 9, 1926, Pressey Papers.

15. Pressey to Remington Typewriter Company, April 9, 1926.

16. Central Scientific Company to Sidney Pressey, April 15, 1926, Pressey Papers; Remington Typewriter Company to Sidney Pressey, April 19, 1926, Pressey Papers.

17. Underwood Typewriter Company to Sidney Pressey, April 15, 1926, Pressey Papers.

18. The Dalton Adding Machine Company to Sidney Pressey, April 22, 1926, Pressey Papers.

19. A. G. Watson, Marietta Apparatus Company to Sidney Pressey, February 22, 1927, Pressey Papers.

20. Leonard Ayres, *Laggards in Our Schools* (New York: Charities Publication Committee, 1909).

21. Leonard Ayres, Cleveland Trust Company to Sidney Pressey, January 28, 1928, Pressey Papers.

22. H. R. Leonard, General Office Equipment Corporation to Sidney Pressey, December 18, 1928, Pressey Papers.

23. Sidney Pressey. "Machine for Intelligent Tests." US Patent No. 1,749,226, filed June 21, 1928, and issued March 4, 1930.

24. Pressey, "A Simple Apparatus," 40.

25. Frederick Winslow Taylor, *The Principles of Scientific Management* (New York: Harper & Brothers, 1911).

26. Raymond Callahan, *Education and the Cult of Efficiency* (Chicago: Chicago University Press, 1962), 23.

27. Ellwood P. Cubberly, *Public School Administration* (Boston: Houghton Mifflin Company, 1916): 337–338.

28. George Brown to Sidney Pressey, February 16, 1926, Pressey Papers.

29. Pressey, "A Simple Apparatus," 40.

30. Pressey, 40.

31. Pressey, 40.

32. Sidney Pressey. "A Machine for Automatic Teaching of Drill Material," *School and Society* 25, no. 645 (May 1927): 549–552.

33. Edward Thorndike, *Animal Intelligence* (New York: The Macmillan Company, 1911).

34. S. A. Courtis to Sidney Pressey, March 30, 1926, Pressey Papers.

35. Sidney Pressey to S. A. Courtis, December 12, 1927, Pressey Papers.

36. "History of Sargent Welch," accessed January 10, 2020, https://www.sargentwelch.com/cms/history_of_sargent_welch.

37. "Diploma Business," *Time*, February 27, 1939, 60.

38. Sidney Pressey to W. M. Welch, W. M. Welch Manufacturing Company, January 9, 1929, Pressey Papers.

39. M. W. Welch to Sidney Pressey, May 14, 1929, Pressey Papers.

40. Welch to Pressey, May 14, 1929.

41. Sidney Pressey to M. W. Welch, June 11, 1929, Pressey Papers.

42. Ellwood P. Cubberly to Sidney Pressey, May 31, 1929, Pressey Papers.

43. Benjamin D. Wood to Sidney Pressey, September 10, 1929, Pressey Papers.

44. W. M. Welch to Sidney Pressey, June 17, 1929, Pressey Papers.

45. Sidney Pressey to M. W. Welch, June 19, 1929, Pressey Papers; Sidney Pressey to R. E. Welch, M. W. Welch Manufacturing Company, July 24, 1929, Pressey Papers; Sidney Pressey to R. E. Welch, September 12, 1929, Pressey Papers; Sidney Pressey to R. E. Welch, September 17, 1929, Pressey Papers.

46. Pressey to Welch, September 17, 1929.

47. Sidney Pressey to R. E. Welch, September 18, 1929, Pressey Papers.

48. R. E. Welch to Sidney Pressey, September 21, 1929, Pressey Papers.

49. Sidney Pressey to R. E. Welch, September 23, 1929, Pressey Papers.

50. R. E. Welch to Sidney Pressey, October 3, 1929, Pressey Papers.

51. Sidney Pressey to R. E. Welch, November 15, 1929, Pressey Papers.

52. R. E. Welch to Sidney Pressey, November 22, 1929, Pressey Papers.

53. Sidney Pressey to R. E. Welch, December 10, 1929, Pressey Papers.

54. George Arps, "Report of the College of Education for the Year Ending June 30, 1930," *Sixtieth Annual Report of the Board of Trustees of The Ohio State University for the Year Ending June 30, 1930* (Columbus: The Ohio State University Press, 1930), 154.

55. R. E. Welch to Luella C. Pressey, February 4, 1930, Pressey Papers.

56. R. E. Welch to Sidney Pressey, February 4, 1930, Pressey Papers.

57. R. E. Welch to Sidney Pressey, April 11, 1930, Pressey Papers.

58. "First Results with, and Problems in the Development of, Apparatus for Testing and Automatic Experimentation in Learning," Pressey Papers.

59. Sidney Pressey to Harry Gilchriese, June 13, 1930, Pressey Papers.

60. "Apparatus Is Invented by Professor to Test, Score and Help in Teaching," *Columbus Dispatch*, April 13, 1930.

61. "Exams by Machinery," *Ohio State University Monthly* (May 1931): 339.

62. M. W. Welch to Sidney Pressey, June 11, 1930, Pressey Papers.

63. Stephen Petrina, "Sidney Pressey and the Automation of Education, 1924–1934," *Technology and Culture* 54, no. 2 (April 2004): 323.

64. Sidney Pressey to M. W. Welch, June 30, 1930, Pressey Papers.

65. R. E. Welch to Sidney Pressey, October 21, 1930, Pressey Papers.

66. R. E. Welch to Sidney Pressey, March 10, 1930, Pressey Papers.

67. R. E. Welch to Sidney Pressey, October 21, 1930, Pressey Papers.

68. Sidney Pressey to M. W. Welch, undated, Pressey Papers.

69. Sidney Pressey to Erwin Esper, January 24, 1941, Pressey Papers.

70. M. W. Welch to Sidney Pressey, August 26, 1931, Pressey Papers.

71. Sidney Pressey to Coin-o-Matic Corporation, October 30, 1931, Pressey Papers; Waldemar Ayres, International Business Machines Corporation to Sidney Pressey, June 5, 1933, Pressey Papers; F. O. Clements, General Motors Corporation to J. L. Morrill, Ohio State University, January 9, 1934, Pressey Papers; B. L. Edholm, Burroughs Adding Machine Company to Sidney Pressey, June 22, 1934, Pressey Papers.

72. Sidney Pressey, "A Third and Fourth Contribution Toward the Coming 'Industrial Revolution' in Education," *School and Society* 36, no. 934 (November 1932): 672.

73. Luella Cole married an anthropology professor; in turn, Pressey married her former best friend, Alice Donnelly, a faculty member in Ohio State University's School of Home Economics Education.

CHAPTER 3

1. Ben D. Wood and Frank N. Freeman, *An Experimental Study of the Educational Influences of the Typewriter in the Elementary School Classroom* (Chicago: The Macmillan Company, 1932).

2. Matthew Downey, *Ben D. Wood: Educational Reformer* (Princeton, NJ: Educational Testing Service, 1965), 49.

3. "Typewriting Aids Students," *San Francisco Examiner*, June 16 1932, 17.

4. Albert Edward Wiggam, "Children at the Typewriter," *School and Society* 49, no. 1260 (1939): 213.

5. Wiggam, "Children at the Typewriter," 214.

6. Educational Bureau of Portable Typewriter Manufacturers, "Notes from 9 July 1928 meeting," Ben D. Wood Papers, Educational Testing Service Archives (hereafter, Wood Papers).

7. Donald Lyman to unknown recipient, November 29, 1932, Wood Papers.

8. Ben D. Wood to Frank Kondolf, September 27, 1929, Wood Papers.

9. Wood and Freeman, *An Experimental Study of the Educational Influences of the Typewriter*, 150.

10. "Eastman Teaching Film Experiment," Wood Papers.

11. Larry Cuban, *Teachers and Machines: The Classroom Use of Technology Since 1920* (New York: Teachers College Press, 1986), 9.

12. Paul Saettler claims that the first regular use of instructional film was in 1910 by the public schools in Rochester, New York. Paul Saettler, *A History of Instructional Technology* (New York: McGraw-Hill, 1968), 98.

13. Cuban, *Teachers and Machines*, 9.

14. Cuban, 9–10.

15. Ben D. Wood, "Criteria of Individualized Education," *Teachers College Record* 38, no. 3 (December 1936): 230.

16. Ben D. Wood, "Mechanical Education Wanted," *Harvard Teachers Record* 1 (1931): 46.

17. This is just one example of how this book tells the particularly *American* version of the story of teaching machines. In West Germany, for example, C. J. Kirsch argues that the machines were used less to further students' individual needs and more to foster a collective exploration of ideas. C. J. Kirsch, "The Germomat III System: Made for the Factory and the Laboratory" (forthcoming).

18. Ben D. Wood to John Dewey, November 22, 1935, cited in Downey, *Ben D. Wood*, 24; John Dewey, *The Child and the Curriculum* (Chicago: University of Chicago Press, 1902).

19. Wood to Dewey, cited in Downey, *Ben D. Wood*, 24.

20. Ben D. Wood, "Continuity in Personnel Work," *News Bulletin of the Bureau of Vocational Information* 4 (March 1926): 17, cited by Downey, *Ben D. Wood*, 26.

21. Ben D. Wood, "Major Strategy versus Minor Tactics in Educational Testing," *Baltimore Bulletin of Education* 13 (September 1934): 7, cited by Downey, *Ben D. Wood*, 27.

22. Downey, *Ben D. Wood*, 8–9.

23. Downey, 13.

24. Downey, 16.

25. Downey, 51.

26. Downey, 51.

27. Downey, 52.

28. "A Mechanical School Teacher," *The Chicago Tribune*, 1933.

29. G. W. Baehne to Superintendent of Schools, October 5, 1933, Wood Papers.

30. Reynold B. Johnson to G. W. Baehne, October 9, 1933, Wood Papers.

31. G. W. Baehne to Reynold B. Johnson, October 11, 1933, Wood Papers.

32. Reynold B. Johnson to IBM, November 3, 1933, Wood Papers.

33. G. W. Baehne to Reynold B. Johnson, November 13, 1933, Wood Papers.

34. Reynold B. Johnson to IBM, November 28, 1933, Wood Papers.

35. J. E. Holt to Reynold B. Johnson, January 2, 1934, Wood Papers.

36. Director of Market Research, IBM, to Reynold B. Johnson, February 8, 1934, Wood Papers.

37. Reynold B. Johnson to J. E. Holt, March 8, 1934, Wood Papers.

38. G. W. Baehne to Reynold B. Johnson, March 20, 1934, Wood Papers.

39. G. W. Baehne to Reynold B. Johnson, May 15, 1934, Wood Papers.

40. Baehne to Johnson, May 15, 1934.

41. G. W. Baehne to G. B. Pegram, July 18, 1934, Wood Papers.

42. G. W. Baehne to Reynold B. Johnson, July 23, 1934, Wood Papers.

43. Reynold B. Johnson to G. W. Baehne, July 26, 1934, Wood Papers. $15,000 is roughly equivalent to $285,000 in 2019 dollars.

44. Johnson to Baehne, July 26, 1934. $4,000 is roughly equivalent to $76,000 in 2019 dollars.

45. Ben D. Wood to Thomas J. Watson, August 4, 1934, Wood Papers.

46. William Blankenship, "Rey Johnson: A Full life, A Fuller Future," *THINK* (June 1971): 40, https://www.ibm.com/ibm/history/exhibits/builders/builders_johnson.html.

47. Downey, *Ben D. Wood*, 53.

48. Downey, 53.

CHAPTER 4

1. Robert Gross to Sherman Fairchild, undated, Papers of Burrhus Frederic Skinner, Harvard University Archives (hereafter, Skinner Papers).

2. H. R. Keith to B. F. Skinner, September 22, 1954, Skinner Papers.

3. B. F. Skinner to H. R. Keith, October 15, 1954, Skinner Papers.

4. B. F. Skinner to P. N. Whittaker, January 11, 1955, Skinner Papers.

5. B. F. Skinner to P. N. Whittaker, August 15, 1955, Skinner Papers.

6. B. F. Skinner to William Reid, November 2, 1955, Skinner Papers.

7. Skinner to Reid, November 2, 1955.

8. B. F. Skinner to P. N. Whittaker, November 28, 1955, Skinner Papers.

9. B. F. Skinner, *A Matter of Consequences* (New York: Alfred A. Knopf, 1983), 70.

10. B. F. Skinner, *The Shaping of a Behaviorist* (New York: New York University Press, 1979), 275.

11. B. F. Skinner, "Baby in a Box—Introducing the Mechanical Baby Tender," *Ladies Home Journal* 62 (October 1945): 30.

12. Skinner, "Baby in a Box," 135.

13. Skinner, 30.

14. Karal Ann Marling, *As Seen on TV: The Visual Culture of Everyday Life in the 1950s* (Cambridge, MA: Harvard University Press, 1994), 255.

15. Alexandra Rutherford, *Beyond the Box: B. F. Skinner's Technology of Behavior from Laboratory to Life, 1950s–1970s* (Toronto: University of Toronto Press, 2009), 26.

16. Nancy McKean to B. F. Skinner, September 29, 1945, Skinner Papers.

17. "A Reader of the Times" to District Attorney, Bloomington, Indiana, September 30, 1945, Skinner Papers.

18. Leila Roosevelt Davis to B. F. Skinner, October 8, 1945, Skinner Papers.

19. Eunice K. Shriver to B. F. Skinner, April 9, 1965, Skinner Papers.

20. B. F. Skinner to Edward L. Thorndike, November 27, 1945, Skinner Papers.

21. B. F. Skinner to Lewis Terman, October 1, 1946, Skinner Papers.

22. J. Weston Judd to B. F. Skinner, undated, Skinner Papers.

23. B. F. Skinner to J. Weston Judd, October 15, 1945, Skinner Papers.

24. J. Weston Judd to B. F. Skinner, October 19, 1945, Skinner Papers.

25. H. J. Glickman to B. F. Skinner, December 6, 1945, Skinner Papers.

26. B. F. Skinner to J. Weston Judd, December 1, 1945, Skinner Papers.

27. B. F. Skinner to J. Weston Judd, April 14, 1946, Skinner Papers.

28. B. F. Skinner to Mrs. H. M. Gibson, April 21, 1946, Skinner Papers.

29. Josephine Schulkins, Cleveland Better Business Bureau, to J. Weston Judd, May 8, 1946, Skinner Papers.

30. B. F. Skinner to Francis Keppel, June 21, 1956, Skinner Papers.

31. Skinner, *A Matter of Consequences*, 118.

32. Skinner, 119.

33. B. F. Skinner to L. E. Bechtel, January 24, 1957, Skinner Papers.

34. B. F. Skinner to L. A. Bechtel, April 29, 1957, Skinner Papers. (Note: some of the letters in Skinner's archives are addressed to L. E. Bechtel and some to L. A. Bechtel.)

35. B. F. Skinner to L. A. Bechtel, June 4, 1957, Skinner Papers.

36. G. W. Youngdale, Jr. to B. F. Skinner, June 14, 1957, Skinner Papers.

37. G. W. Youngdale, Jr. to B. F. Skinner, August 27, 1957, Skinner Papers.

38. Skinner, *A Matter of Consequences*, 137.

39. Wayne Urban, *More Than Science and Sputnik: The National Defense Education Act of 1958* (Tuscaloosa: University of Alabama Press, 2010), 55.

40. Urban, *More Science Than Sputnik*, 80.

41. Urban, 81.

42. Hyman Rickover, *Education and Freedom* (New York: Dutton, 1959).

43. David A. Loehwing, "Aids to Education: Teaching Devices Are Proving Their Usefulness in U.S. Schools," *Barron's National Business and Financial Weekly*, May 16, 1960, 3, 18.

44. Loehwing, "Aids to Education," 18.

45. Daniel Bjork, *B. F. Skinner: A Life* (New York: Basic Books, 1993), 177.

46. Bjork, *B. F. Skinner*, 177.

47. William Jovanovich to B. F. Skinner, April 16, 1958, Skinner Papers.

48. William Jovanovich to B. F. Skinner, April 17, 1958, Skinner Papers.

49. B. F. Skinner to J. H. McCallum, June 26, 1958, Skinner Papers.

50. J. H. McCallum, "Memorandum," July 15, 1958, Skinner Papers.

51. J. H. McCallum to B. F. Skinner, July 16, 1958, Skinner Papers.

52. J. H. McCallum to B. F. Skinner, July 25, 1958, Skinner Papers.

53. J. H. McCallum, "Memorandum," July 30, 1958, Skinner Papers.

54. Paul F. Perkins, Jr. to B. F. Skinner, August 1, 1958, Skinner Papers.

55. William Jovanovich to Charles Argulo, August 6, 1958, Skinner Papers.

56. J. H. McCallum to B. F. Skinner, August 7, 1958, Skinner Papers.

57. J. H. McCallum to B. F. Skinner, September 16, 1958, Skinner Papers.

58. R. E. Zenner to B. F. Skinner, October 20, 1958, Skinner Papers.

59. B. F. Skinner to J. H. McCallum, October 30, 1958, Skinner Papers. There are, in fact, two versions of this letter from Skinner to McCallum in Skinner's archives. One was likely written for McCallum's use in contacting Comptometer's Peter Mero.

60. J. H. McCallum to Peter Mero, November 7, 1958, Skinner Papers.

61. Skinner, *A Matter of Consequences*, 144.

62. B. F. Skinner to J. H. McCallum, November 17, 1958, Skinner Papers.

63. H. W. Miller, Jr. to B. F. Skinner, December 31, 1958, Skinner Papers.

64. B. F. Skinner to H. W. Miller, Jr., January 5, 1959, Skinner Papers.

65. H. W. Miller, Jr. to B. F. Skinner, February 4, 1959, Skinner Papers.

66. Skinner, *A Matter of Consequences*, 144.

67. B. F. Skinner to Harold Robson, June 19, 1959, Skinner Papers.

68. B. F. Skinner to Donald Rivkin, July 6, 1959, Skinner Papers.

CHAPTER 5

1. B. F. Skinner, *A Matter of Consequences* (New York: Alfred A. Knopf, 1983), 158.

2. Skinner, *A Matter of Consequences*, 158.

3. B. F. Skinner to Dean Luxton, July 20, 1959, Papers of Burrhus Frederic Skinner, Harvard University Archives (hereafter, Skinner Papers).

4. S. S. Stevens to B. F. Skinner, August 7, 1959, Skinner Papers.

5. Edwin B. Newman to B. F. Skinner, July 29, 1959, Skinner Papers.

6. Walter Lewis to B. F. Skinner, August 13, 1959, Skinner Papers.

7. B. F. Skinner to Dean Luxton, August 21, 1959, Skinner Papers.

8. B. F. Skinner to Walter Lewis, August 26, 1959, Skinner Papers.

9. Emma Harrison, "Teacher Machine to Be Ready in '61," *New York Times,* September 6, 1959, 61.

10. B. F. Skinner to Dean Luxton, Donald Burdorf, and C. E. Philips, September 10, 1959, Skinner Papers.

11. B. F. Skinner to Donald Burdorf, September 10, 1959, Skinner Papers.

12. B. F. Skinner to Dean Luxton, September 15, 1959, Skinner Papers.

13. B. F. Skinner to Dean Luxton, September 18, 1959, Skinner Papers.

14. Donald Rivkin to B. F. Skinner, September 21, 1959, Skinner Papers.

15. B. F. Skinner to Donald Rivkin, October 14, 1959, Skinner Papers.

16. "Rheem Acquires Califone Corp," *The Independent* (Long Beach, CA), October 22, 1959, 68. One million dollars is roughly the equivalent of $8 million today.

17. B. F. Skinner to A. Lightfoot Walker, October 16, 1959, Skinner Papers.

18. Dean Luxton to B. F. Skinner, November 4, 1959, Skinner Papers.

19. Robert Metzner to B. F. Skinner, November 6, 1959, Skinner Papers.

20. B. F. Skinner to Robert G. Metzner, December 11, 1959, Skinner Papers.

21. Robert Metzner to B. F. Skinner, December 18, 1959, Skinner Papers.

22. B. F. Skinner to Donald Burdorf, February 10, 1960, Skinner Papers; B. F. Skinner to Donald Burdorf, February 19, 1960, Skinner Papers.

23. B. F. Skinner to Donald Rivkin, February 15, 1960, Skinner Papers.

24. B. F. Skinner to Donald Burdorf, February 24, 1960, Skinner Papers.

25. B. F. Skinner to I. G. Davis, March 8, 1960, Skinner Papers.

26. Skinner to Davis, March 8, 1960.

27. John W. Gardner, foreword to *The American High School Today* by James Conant Bryant (New York: McGraw-Hill, 1959), ix, x–xi.

28. James B. Conant, *The American High School Today* (New York: McGraw-Hill, 1959), 15.

29. Conant, *The American High School Today*, 12.

30. Conant, 14.

31. James B. Conant, *Shaping Educational Policy* (New York: McGraw-Hill, 1964), 40.

32. Conant, *The American High School Today*, 40.

33. Fred M. Hechinger, "Most Bright Pupils Are Not Working Hard Enough," *New York Times*, February 15, 1959, 125.

34. "The American High School Today," *Hispania* 42, no. 1 (March 1959): 155.

35. David Tyack and Larry Cuban, *Tinkering Toward Utopia: A Century of Public School Reform* (Cambridge, MA: Harvard University Press, 1995), 48.

36. Audrey Watters, "The Invented History of 'The Factory Model of Education,'" *Hack Education*, April 25, 2015, http://hackeducation.com/2015/04/25/factory-model.

37. Conant, *The American High School Today*, 6–7.

38. Spencer Klaw, "What Can We Learn from Teaching Machines?," *Reporter* 27, no. 2 (July 19, 1962): 22.

39. B. F. Skinner, "Teaching Machines," *Science* 128, no. 3330 (October 24, 1958).

40. Skinner, *A Matter of Consequences*, 166.

41. B. F. Skinner to James B. Conant, December 16, 1959, Skinner Papers.

42. Skinner, *A Matter of Consequences*, 167.

43. Skinner, 167.

44. B. F. Skinner, *The Shaping of a Behaviorist* (New York: New York University Press, 1979), 131.

45. Daniel W. Bjork, *B. F. Skinner: A Life* (New York: Basic Books, 1993), 184.

46. John W. Gardner, foreword to Conant, *The American High School Today*, xii.

47. Skinner, *A Matter of Consequences*, 166.

48. Skinner, 167–168. B. F. Skinner to I. G. Davis, January 4, 1960, Skinner Papers.

49. B. F. Skinner to James B. Conant, January 14, 1960, Skinner Papers.

50. Bjork, *B. F. Skinner*, 185.

51. James B. Conant to B. F. Skinner, January 18, 1960, Skinner Papers.

52. Conant to Skinner, January 18, 1960.

53. B. F. Skinner to James B. Conant, February 4, 1960, Skinner Papers.

54. B. F. Skinner to I. G. Davis, February 11, 1960, Skinner Papers.

55. B. F. Skinner to I. G. Davis, January 12, 1960, Skinner Papers.

56. B. F. Skinner to Webster Jones, February 5, 1960, Skinner Papers.

57. B. F. Skinner to I. G. Davis, April 18, 1960, Skinner Papers.

58. I. G. Davis to B. F. Skinner, May 6, 1960, Skinner Papers.

59. B. F. Skinner to Donald Rivkin, May 12, 1960, Skinner Papers.

60. Donald Rivkin to B. F. Skinner, May 13, 1960, Skinner Papers.

61. B. F. Skinner to I. G. Davis, May 31, 1960, Skinner Papers.

62. B. F. Skinner to Donald Rivkin, July 5, 1960, Skinner Papers.

63. B. F. Skinner to Donald Rivkin, July 15, 1960, Skinner Papers.

64. B F. Skinner to Donald Rivkin, July 28, 1960, Skinner Papers.

65. Donald Rivkin to B. F. Skinner, August 11, 1960, Skinner Papers.

66. Gene Bylinsky, "Robot Teachers: Schools, Business Firms Spur Use of Machines to Drill, Test Students," *Wall Street Journal*, August 8, 1960, 1.

67. B. F. Skinner to Donald Rivkin, September 22, 1960, Skinner Papers.

68. B. F. Skinner to O. X. Pitney, October 21, 1960, Skinner Papers.

CHAPTER 6

1. James G. Holland would join the team two years later.

2. Daniel W. Bjork, *B. F. Skinner: A Life* (New York: Basic Books, 1993), 174–175.

3. B. F. Skinner, *A Matter of Consequences* (New York: Alfred A. Knopf, 1983), 119.

4. According to Daniel Tröhler, the phrase "programmed instruction" was first explained in a 1955 pamphlet by the US Air Force's Guy Besnard, Leslie Briggs, George Mursch, and Elbert Walker—"The Development of the Subject-Matter Trainer." Briggs, an Air Force psychologist, had been a doctoral student of Sidney Pressey. Daniel Tröhler, "The Technocratic Momentum after 1945, the Development of Teaching Machines, and Sobering Results," *Journal of Educational Media, Memory, and Society* 5, no. 2 (Autumn 2013): 10–11.

5. Skinner, *A Matter of Consequences*, 120.

6. James G. Holland, "A Teaching Machine Program in Psychology," in *Automatic Teaching: The State of the Art*, ed. Eugene Galanter (New York: John Wiley & Sons, 1959), 69.

7. Bjork, *B. F. Skinner*, 175. Bjork cites the *Harvard Alumni Bulletin*, May 1960, 638.

8. Susan Meyer Markle was married three times: first to Arthur Meyer, then to David Markle (shortly after her arrival at Harvard), and finally to Phil Tiemann. As Dale Brethower writes in a memorial, "Many years later, having earned an excellent and thoroughly well-deserved reputation in the field of performance improvement/programmed learning, she was asked to offer advice to others in the field, especially young women. 'Always publish under your maiden name!' was the advice. She had developed a solid professional reputation, first as 'Susan Meyer' and then again as 'Susan Markle.' Because no one had told her to 'Always publish under your maiden name!' she was unable to follow the sage advice herself." (Although she did not change her name professionally upon her third marriage.) Dale M. Brethower, "In Memoriam: Susan Meyer Markle (1928–2008)," *Performance Improvement* 50, no. 1 (January 2011): 6.

9. Skinner, *A Matter of Consequences*, 119.

10. Susan Meyer Markle to B. F. Skinner, April 24, 1979, The Papers of Burrhus Frederic Skinner, Harvard University Archives (hereafter, Skinner Papers).

11. "Fading," for example, is the process of slowly removing prompts in the programming materials as a student begins to demonstrate that they need fewer clues. These clues or cues are also known as prompts, and "prompting" was one technique to help minimize incorrect answers.

12. Susan Meyer Markle, Lewis D. Eigen, and P. Kenneth Komoski, *A Programed Primer on Programing* (New York: The Center for Programed Instruction, Inc., 1961); Susan Meyer Markle, *Good Frames and Bad: A Grammar of Frame Writing* (New York: John Wiley & Sons, 1964).

13. Susan Meyer Markle, "Inside the Teaching Machine," *Saturday Review*, November 18, 1961, 55.

14. Susan Meyer Markle, "Individualizing Programed Instruction: The Programer's Part," in *Programs, Teachers, and Machines*, ed. Alfred de Grazia and David A. Sohn (New York: Bantam Books, 1964), 146.

15. Meyer Markle, "Individualizing Programed Instruction," 146.

16. Norman Crowder, *The Arithmetic of Computers* (New York: Doubleday, 1958).

17. Advertisement, *Popular Science*, July 1961, 10.

18. Crowder, *The Arithmetic of Computers*, v.

19. Crowder, v.

20. Norman Crowder, "On the Differences between Linear and Intrinsic Programing," *The Phil Delta Kappan* 44, no. 6 (March 1963): 253. For more on the development of the Choose Your Own Adventure series, see Jake Rossen, "A Brief History of 'Choose Your Own Adventure,'" *Mental Floss*, April 10, 2014, http://mentalfloss.com/article/56160/brief-history-choose-your-own-adventure. Eli Cook, "Rearing Children of the Market in the 'You' Decade: Choose Your Own Adventure Books and the Ascent of Free Choice in 1980s America," *Journal of American Studies* (February 2020): 1–28.

21. Crowder, "On the Differences between Linear and Intrinsic Programing," 250.

22. Norman Crowder, "Automatic Tutoring by Intrinsic Programming," in *Teaching Machines and Programmed Learning: A Source Book*, ed. A. A. Lumsdaine and Robert Glaser (Washington, DC: National Education Association, 1960), 286.

23. Crowder, "Automatic Tutoring by Intrinsic Programming," 286.

24. Spencer Klaw, "What Can We Learn from Teaching Machines?," *The Reporter* 27, no. 2 (July 19, 1962): 21.

25. Crowder, "Automatic Tutoring by Intrinsic Programming," 288.

26. See, for example, Benjamin Fine, *Teaching Machines* (New York: Sterling Press, 1962); George A. W. Boehm, "Can People Be Taught Like Pigeons?," *Fortune*, October 1960, 176–178.

27. Meyer Markle, "Individualizing Programed Instruction," 153.

28. Susan Meyer Markle, "The Harvard Teaching Machine Project: The First Hundred Days," *AV Communication Review* 12, no. 3 (Fall 1964): 345.

29. J. M. Reid, Harcourt Brace, to E. P. Smith, Memorandum "Re: Michigan Conference on the Skinner Procedure," July 31, 1958, Skinner Papers.

30. Sidney Pressey, "Sidney Leavitt Pressey," in *A History of Psychology in Autobiography,* ed. Edwin Boring and Gardner Lindzey (New York: Appleton-Century-Crofts, 1967), 332–333.

31. A. A. Lumsdaine and Robert Glaser, "Purpose and Scope of This Book," in *Teaching Machines and Programmed Learning: A Source Book*, ed. A. A. Lumsdaine and Robert Glaser (Washington, DC: National Education Association, 1960), 1–2.

32. Eugene Galantar, ed., *Automatic Teaching: The State of the Art* (New York: John Wiley & Sons, 1959).

33. Lumsdaine and Glaser, ed., *Teaching Machines and Programmed Learning*, 6.

34. Lumsdaine and Glaser, 17.

35. Sidney Pressey, "Certain Major Psycho-Educational Issues Appearing in the Conference on Teaching Machines," in *Automatic Teaching: The State of the Art*, ed. Eugene Galanter (New York: John Wiley & Sons, 1959), 196.

36. Sidney Pressey to Ben D. Wood, February 10, 1962, Sidney Leavitt Pressey Papers, University Archives, Ohio State University Libraries.

37. Pressey, "Sidney Leavitt Pressey," 332–333.

CHAPTER 7

1. Sidney Pressey, "A Third and Fourth Contribution Toward the Coming 'Industrial Revolution' in Education," *School and Society* 36, no. 934 (November 19, 1932): 668.

2. Frederick James Smith, "Looking into the Future with Thomas A. Edison," *New York Dramatic Mirror*, July 9, 1913, 24.

3. Simon Ramo, "A New Technique of Education," in *Teaching Machines and Programmed Learning: A Source Book*, ed. A. A. Lumsdaine and Robert Glaser (Washington, DC: National Education Association, 1960), 367. The Ramo article first appeared in *Engineering and Science* 21, no. 1, 1957. Excerpts used with permission from Caltech.

4. Ramo, "A New Technique of Education," 369.

5. Ramo, 369.

6. Ramo, 369.

7. Ramo, 370.

8. Smith, "Looking into the Future with Thomas A. Edison," 24.

9. Ramo, "A New Technique of Education," 371.

10. Ramo, 371.

11. Ramo, 371.

12. Ramo, 372.

13. Ramo, 372–373.

14. Ramo, 373.

15. Ramo, 373–374.

16. Ramo, 374.

17. Ramo, 372.

18. Ramo, 379.

19. Herbert Simon, "The Job of a College President," *Educational Record* 48 (Winter 1967): 77.

20. Ramo, "A New Technique of Education," 378.

21. Ramo, 379.

22. Raynold Abrashkin and Jay Williams, *Danny Dunn and the Homework Machine* (New York: McGraw-Hill, 1958).

23. Arthur Radebaugh, "Push Button Education," *Closer Than We Think* (syndicated comic), May 25, 1958.

24. Matt Novak, "Before the Jetsons, Arthur Radebaugh Illustrated the Future," *Smithsonian Magazine*, April 2012, https://www.smithsonianmag.com/science-nature/before-the-jetsons-arthur-radebaugh-illustrated-the-future-122729342/.

25. John E. Coulson, ed., *Programmed Learning and Computer-Based Instruction* (New York: John Wiley and Son, 1962); Alfred de Grazia and David A. Sohn, ed., *Programs, Teachers, and Machines* (New York: Bantam Books, 1964); William Deterline, *An Introduction to Programed Instruction* (Englewood Cliffs, NJ: Prentice-Hall, 1962); Sam and Beryl Epstein, *The First Book of Teaching Machines* (New York: Franklin Watts, 1961); Benjamin Fine, *Teaching Machines* (New York: Sterling Press, 1962); Wilbur Schramm, *Programed Instruction: Today and Tomorrow* (New York: Fund for the Advancement of Education, 1962); Edward Fry, *Teaching Machines and Programed Instruction: An Introduction* (New York: McGraw-Hill, 1963); Robert Glaser, ed., *Teaching Machines and Programed Learning II: Data and Directions* (Washington, DC: National Education Association, 1965); Richard Goodman, *Programmed Learning and Teaching Machines: An Introduction* (London: The English Universities Press, 1962); Joseph Roucek, ed., *Programmed Teaching: A Symposium on Automation in Teaching* (New York: Philosophical Library, 1965); E. W. Rushton, *Programmed Learning: The Roanoke Experiment* (Chicago: Encyclopedia Britannica Press, 1965); Lawrence Stolurow, *Teaching by Machine* (Washington, DC: US Department of Health, Education, and Welfare, 1961).

26. Joseph Bell, "Will Robots Teach Your Children?," *Popular Mechanics*, October 1961, 157.

27. C. P. Gilmore, "Teaching Machines—Do They or Don't They?," *Popular Science* 161, no. 6 (December 1962): 166.

28. Spencer Klaw, "What Can We Learn from Teaching Machines?," *The Reporter* 27, no. 2 (July 19, 1962): 20–21.

29. Raymond Callahan, *Education and the Cult of Efficiency* (Chicago: University of Chicago Press, 1962).

30. Daniel Seligman, "The Low Productivity of the 'Education Industry,'" *Fortune*, October 1958, 135–138, 195–196.

31. Seligman, "The Low Productivity of the 'Education Industry,'" 135.

32. Seligman, 135.

33. Seligman, 138.

34. Seligman, 138.

35. Seligman, 195.

36. George A. W. Boehm, "Can People Be Taught Like Pigeons?," *Fortune*, October 1960, 176–178.

37. Boehm, "Can People Be Taught Like Pigeons?," 177.

38. Boehm, 178.

39. B. F. Skinner, "The Science of Learning and the Art of Teaching," *Harvard Educational Review* 24 (1954): 90.

CHAPTER 8

1. G. K. Hodenfield, "New Teaching Machine Plan May Sweep Nation," *Daily Press* (Newport News, VA), October 27, 1960, 28.

2. Bill Lawrence, "Teaching Machines Speed Instruction in Roanoke," Associated Press, *The Progress Index* (Petersburg-Colonial Heights, VA), April 17, 1960, 2.

3. Efforts to desegregate the schools in Roanoke occurred simultaneously to the teaching machine experiment there, although the racial composition of students who participated in the high-profile pilot study is rarely, if ever, remarked upon. For more on school segregation in the community, see Peter Carr Jones, "Integrating 'the Star City of the South': Roanoke School Desegregation and the Politics of Delay," MA thesis, College of William and Mary, 2013.

4. E. W. Rushton, *Programmed Learning: The Roanoke Experiment* (Chicago: Encyclopedia Britannica Press, 1965), 5–6.

5. Rushton, *Programmed Learning*, 6.

6. B. F. Skinner, *A Matter of Consequences* (New York: Alfred Knopf, 1983), 185.

7. Rushton, *Programmed Learning*, 8.

8. Skinner, *A Matter of Consequences*, 185.

9. Rushton, *Programmed Learning*, 9.

10. Rushton, 9.

11. Rushton, 10–11.

12. Rushton, 37.

13. Rushton, 17.

14. "Group Hears Report on Teaching Machines," *The Bee* (Danville, VA), February 15, 1961, 30.

15. Rushton, *Programmed Learning*, 17.

16. Rushton, 17.

17. Rushton, 18–19.

18. Rushton, 19.

19. Rushton, 23.

20. Rushton, 26.

21. Lincoln F. Hanson and P. Kenneth Komoski, "School Use of Programed Instruction," in *Teaching Machines and Programed Learning, II: Data and Directions*, ed. Robert Glaser (Washington, DC: National Education Association, 1965), 654.

22. Hanson and Komoski, "School Use of Programed Instruction," 655.

23. Rushton, *Programmed Learning*, 34.

24. Rushton, 1.

25. Rushton, 35.

26. Rushton, 1.

27. Carr Jones, "Integrating 'the Star City of the South,'" 6.

28. Paul Saettler, *The Evolution of American Educational Technology* (New York: McGraw-Hill, 1990), 433.

29. Associated Press, "Hollins College to End 'Programmed Teaching,'" *Daily Press* (Newport News, VA), December 9, 1960, 35.

30. Sullivan's readers should not be confused with the Harper Collins books with the same name—a series that first appeared in 1957 and included Syd Hoff's beloved *Danny and the Dinosaur* (1958).

31. Lorna Thompson, Jr., "The Sullivan Reading Program, Developed by Sullivan Associates, Menlo Park, California," Contract no. OEC-0-70-4892 (Palo Alto, CA: American Institutes for Research in the Behavioral Sciences, 1971), 20.

32. Thompson, Jr., "The Sullivan Reading Program," 21.

33. Thompson, Jr., 23.

34. Thompson, Jr., 24.

CHAPTER 9

1. "The Truth about Those Teaching Machines," *Changing Times* 16, no. 2 (February 1962): 15.

2. Spencer Klaw, "What Can We Learn from Teaching Machines?," *Reporter* 27, no. 2 (July 19, 1962): 22.

3. Klaw, "What Can We Learn from Teaching Machines?," 22–23.

4. Advertisement placed by the Univox Institute, *New York Times*, March 25, 1962, 21. Second source unknown, 1963, Papers of Burrhus Frederic Skinner, Harvard University Archives (hereafter, Skinner Papers).

5. Yes, this is the James Evans that Susan Meyer Markle said tried to claim that he'd been the first to compete a dissertation on programmed instruction.

6. Klaw, "What Can We Learn from Teaching Machines?," 22.

7. B. F. Skinner to Theodore Waller, February 15, 1961, Skinner Papers.

8. Theodore Waller to B. F. Skinner, March 23, 1967, Skinner Papers.

9. Donald Rivkin to B. F. Skinner, July 19, 1961, Skinner Papers.

10. B. F. Skinner to William Spaulding, July 24, 1961, Skinner Papers.

11. Klaw, "What Can We Learn from Teaching Machines?," 22.

12. Nicholas Gilmore, "Death of a Sales Scheme: Encyclopedia Shysters of the Door-to-Door Age," *Saturday Evening Post*, August 30, 2017, https://www.saturdayeveningpost.com/2017/08/death-sales-scheme-encyclopedia-shysters-door-door-age/.

13. Warner Olivier, "160 Miles of Words," *Saturday Evening Post*, July 21, 1945, 85.

14. James Coleman, *Equality of Educational Opportunity* (Washington, DC: US Department of Health, Education, and Welfare, 1966), 2.

15. Coleman, *Equality of Educational Opportunity*, 325.

16. Coleman, 20.

17. Coleman, 189.

18. "How to Pick an Encyclopedia," *Changing Times* 16, no. 6 (June 1, 1962), 15.

19. "Income of Persons and Families in the United States: 1962," in *Consumer Income* (Washington, DC: Bureau of the Census, 1963), 1.

20. B. F. Skinner, *A Matter of Consequences* (New York: Alfred A. Knopf, Inc., 1983), 37.

21. Felix F. Kopstein and Isabel J. Shillestad, *A Survey of Auto-Instructional Devices* (Arlington, VA: Armed Services Technical Information Agency, September 1961), 28.

22. L. B. Wyckoff, Jr., "Teaching Machines," Patent No. US3137948, filed August 17, 1960 and issued June 23, 1964, 2.

23. C. J. Donnelly, "Problems in Publishing Programmed Materials," in *Trends in Programmed Instruction: Papers from the First Annual Convention of the National Society for Programmed Instruction*, ed. Gabriel By-Ofiesh and Wesley Meierhenny (Washington, DC: National Education Association, 1964), 269.

24. A. A. Lumsdaine, "Some Issues Concerning Devices and Programs for Automated Learning," in *Teaching Machines and Programmed Learning: A Source Book*, ed. A. A. Lumsdaine and Robert Glaser (Washington, DC: National Education Association, 1960), 526.

25. Rudolf Flesh, *Why Johnny Can't Read: And What You Can Do About It* (New York: Harper & Brothers, 1955).

26. Benjamin Fine, *Teaching Machines* (New York: Sterling Press, 1962), 78.

27. B. F. Skinner to Mrs. P. S. Madden, November 8, 1968, Skinner Papers.

28. Skinner, *A Matter of Consequences*, 250–251.

29. "Westinghouse Opens Office for Programmed Learning," *Albuquerque Journal*, May 4, 1965, 2.

30. Rogelio Escobar and Kennon A. Lattal, "Observing Ben Wyckoff: From Basic Research to Programmed Instruction and Social Issues," *The Behavior Analyst* 34, no. 2 (Fall 2001): 164.

31. Wilson Cliff, "Grolier Society Bids to Acquire Assets of TMI," *Albuquerque Journal*, May 26, 1967, 1, 4.

32. Cliff, "Grolier Society Bids to Acquire Assets of TMI," 4.

33. Klaw, "What Can We Learn from Teaching Machines?," 23.

34. Emma Harrison, "Teacher Machine to Be Ready in '61," *New York Times*, September 6, 1959, 61; Fred Hechinger, "Teacher Machine to Be Ready in '61," *New York Times*, June 9, 1960, 25.

35. Edward Summers, "Programed Instruction and the Teaching of Reading," *Teaching Aids News* 5, no. 10 (May 30, 1965): 1.

36. Klaw, "What Can We Learn from Teaching Machines?," 23.

37. Advertisement, *Chicago Daily Tribune*, January 9, 1961, 39.

CHAPTER 10

1. Fred M. Hechinger, "Teacher Machine to Be Ready in '61," *New York Times*, June 9, 1960, 25.

2. B. F. Skinner to Walter S. Lewis, June 23, 1960, The Papers of Burrhus Frederic Skinner, Harvard University Archives (hereafter, Skinner Papers).

3. B. F. Skinner to O. X. Pitney, October 21, 1960, Skinner Papers.

4. B. F. Skinner to O. X. Pitney, December 8, 1960, Skinner Papers.

5. Harry R. Sage to Daniel A. Austin, Jr., February 20, 1961, Skinner Papers.

6. B. F. Skinner to C. V. Coons, April 11, 1961, Skinner Papers.

7. C. V. Coons to B. F. Skinner, April 17, 1961, Skinner Papers.

8. C. V. Coons to B. F. Skinner, May 16, 1961, Skinner Papers.

9. B. F. Skinner to O. X. Pitney, June 1, 1961, Skinner Papers.

10. C. V. Coons to B. F. Skinner, June 2, 1961, Skinner Papers.

11. For more on Gordon Pask, see Gordon Pask, *An Approach to Cybernetics* (London: Hutchinson & Co, 1961); Pask, "Machines That Teach," *New Scientist*, no. 234 (May 11, 1961), 308–311; Pask, *The Cybernetics of Human Learning and Performance* (London: Hutchinson & Co, 1975).

12. Donald Rivkin to B. F. Skinner, June 14, 1961, Skinner Papers.

13. B. F. Skinner to C. V. Coons, June 26, 1961, Skinner Papers.

14. C. V. Coons to B. F. Skinner, July 31, 1961, Skinner Papers.

15. Daniel Bjork, *B. F. Skinner: A Life* (New York: Basic Books, 1993), 181. Here, Bjork is citing a note from the Skinner basement archives titled "Stock Taking" and dated April 21, 1961.

16. B. F. Skinner to C. V. Coons, September 7, 1961, Skinner Papers.

17. "Publication Rights to Teaching Machine Programs owned by Arthur C. Croft Publications," September 11, 1961, Skinner Papers.

18. William Spaulding, "To the Record: Meeting with Messrs. Padwa, Simmons and Baldwin of Basic Systems Inc., September 11, 1961," September 12, 1961, Skinner Papers.

19. William Spaulding to B. F. Skinner, January 23, 1962, Skinner Papers; James G. Holland and B. F. Skinner, *The Analysis of Behavior* (New York: McGraw-Hill, 1961).

20. "Raytheon to Buy a Unit of Rheem," *New York Times*, October 25, 1961, 51.

21. Teaching Machine timeline, Skinner Papers.

22. B. F. Skinner to C. V. Coons, December 27, 1961, Skinner Papers.

23. B. F. Skinner to C. V. Coons, February 9, 1962, Skinner Papers.

24. G. W. Mallatratt to B. F. Skinner, February 13, 1962, Skinner Papers.

25. B. F. Skinner to George Mallatratt, February 20, 1962, Skinner Papers.

26. G. W. Mallatratt to B. F. Skinner, March 26, 1962, Skinner Papers.

27. B. F. Skinner to G. W. Mallatratt, March 28, 1962, Skinner Papers.

28. G. W. Mallatratt to B. F. Skinner, April 6, 1962, Skinner Papers.

29. B. F. Skinner to G. W. Mallatratt, April 9, 1962, Skinner Papers.

30. B. F. Skinner to G. W. Mallatratt, June 6, 1962, Skinner Papers.

31. A. J. Budden to B. F. Skinner, September 24, 1962, Skinner Papers.

32. B. F. Skinner to G. W. Mallatratt, November 12, 1962, Skinner Papers.

33. B. F. Skinner to Walter Lewis, January 23, 1963, Skinner Papers.

34. B. F. Skinner to G. W. Mallatratt, February 26, 1963, Skinner Papers.

35. B. F. Skinner to G. W. Mallatratt, March 28, 1963, Skinner Papers.

36. B. F. Skinner, "Business," November 20, 1962, Skinner basement archives, as cited in Bjork's *B. F. Skinner*, 181.

37. Donald Rivkin to B. F. Skinner, September 12, 1963, Skinner Papers.

CHAPTER 11

1. B. F. Skinner, *The Shaping of a Behaviorist* (New York: New York University Press, 1984), 292.

2. B. F. Skinner to Toni Knoss, July 13, 1964, The Papers of Burrhus Frederic Skinner, Harvard University Archives (hereafter, Skinner Papers).

3. B. F. Skinner, *Walden Two* (New York: Macmillan, 1948).

4. Skinner, *The Shaping of a Behaviorist*, 296.

5. Skinner, 296.

6. Daniel Bjork, *B. F. Skinner: A Life* (New York: Basic Books, 1993), 149.

7. B. F. Skinner, "Walden Two Revisited," *Walden Two* (Indianapolis: Hackett Publishing Company, [1948] 1975), iv.

8. Deborah E. Altus and Edward K. Morris, "B. F. Skinner's Utopian Vision: Behind and Beyond *Walden Two*," *The Behavior Analyst* 32, no. 2 (Fall 2009): 320; B. F. Skinner, *The Behavior of Organisms: An Experimental Analysis* (New York: Appleton-Century, 1938).

9. B. F. Skinner, *A Matter of Consequences* (New York: Alfred Knopf, 1983), 252.

10. Bjork, *B. F. Skinner*, 162.

11. Skinner, "Walden Two Revisited," vii.

12. Skinner, vii.

13. A. S. Neill, *Summerhill: A Radical Approach to Child Rearing* (New York: Hart Publishing Company, 1960), 162.

14. Jonathan Croall, *Neill of Summerhill: The Permanent Rebel* (New York: Pantheon Books, 1983), 353.

15. George Dennison, "Freedom to Grow," *New York Times*, October 16, 1966, 329–330.

16. Paul Goodman, *Growing Up Absurd* (New York: Vintage Books, 1960), 48.

17. Goodman, *Growing Up Absurd*, 122–123.

18. For more on Goodman and the politics of Gestalt therapy, see Jack Aylward, "The Contributions of Paul Goodman to the Clinical, Social, and Political Implications of Boundary Disturbances," *Gestalt Review* 3, no. 2 (1999): 107–118.

19. Paul Goodman, *Compulsory Mis-education* (New York: Vintage Books, 1964), 18.

20. Goodman, *Compulsory Mis-education*, 28.

21. Goodman, 29.

22. Goodman, 59.

23. Goodman, 10.

24. Goodman, 145.

25. Goodman, 153.

26. Goodman, 60.

27. Goodman, 60–61.

28. Goodman, 90.

29. Eric Burner, *And Gently He Shall Lead Them: Robert Parris Moses and Civil Rights in Mississippi* (New York: New York University Press, 1994), 91.

30. Among the funders was the Taconic Foundation, whose support was secured by Burke Marshall, the head of the Civil Rights Division of the Department of Justice, whose involvement "underscores the point that there was a closer relationship between SNCC and federal officials than hitherto recognized," writes Laura Visser-Maessen in *Robert Parris Moses, A Life in Civil Rights and Leadership at the Grassroots* (Durham: University of North Carolina Press, 2016), loc. 3147–3148.

31. Burner, *And Gently He Shall Lead Them*, 92.

32. "Delta Ministry Fact Sheet," National Council of Churches Delta Ministry, January 1965, appendix D.

33. Noel Day, cited by Mary Aickin Rothschild in "The Volunteers and the Freedom Schools: Education for Social Change in Mississippi," *History of Education Quarterly* 22, no. 4 (Winter 1982): 406.

34. John Gregory Speed, "A Beacon of Light: Tougaloo During the Presidency of Dr. Adam Daniel Beittel (1960–1964)," PhD diss., the University of Southern Mississippi, 2014, 260.

35. Speed, "Beacon of Light," 261.

36. Rogelio Escobar and Kennon A. Lattal, "Observing Ben Wyckoff: From Basic Research to Programmed Instruction and Social Issues," *The Behavior Analyst* 34, no. 2 (Fall 2001): 38.

37. Len Holt, *The Summer That Didn't End: The Story of the Mississippi Civil Rights Project of 1964* (New York: De Capo Press, 1965): 152–153.

38. Holt, *The Summer That Didn't End*, 153.

39. Daniel Perlstein, "Teaching Freedom: SNCC and the Creation of the Mississippi Freedom Schools," *History of Education Quarterly* 30, no. 3 (Autumn 1999): 304.

40. John R. Rachal, "We'll Never Turn Back: Adult Education and the Struggle for Citizenship in Mississippi's Freedom Summer," *American Educational Research Journal* 35, no. 2 (Summer 1998): 170.

41. Paulo Freire, *Pedagogy of the Oppressed* (New York: The Continuum International Publishing Group, 2005), 72.

42. Goodman, *Compulsory Mis-education*, 80.

43. Goodman, 80.

44. Goodman, 88.

45. Perlstein, "Teaching Freedom," 304. Perlstein is citing Charles Silberman's definition of programmed learning from *Crisis in the Classroom: The Remaking of American Education* (New York: Random House, 1970), 196–197.

46. Daniel Perlstein, from an interview with Robert Parris Moses, "Teaching Freedom," 306. There was, to be fair, commercial interest in the materials that John Blyth had developed for Moses, and The Diebold Group took Blyth to court for ownership of them.

47. Goodman, *Compulsory Mis-education*, 88.

48. Goodman, 89.

49. Freire, *Pedagogy of the Oppressed*, 133.

50. Goodman, *Compulsory Mis-education*, 90–91.

51. Seth Rosenfeld, *Subversives: The FBI's War on Student Radicals, and Reagan's Rise to Power* (London: Macmillan, 2012), 216–217.

52. Steven Lubar, "'Do Not Fold, Spindle, or Mutilate': A Cultural History of the Punch Card," *Journal of American Culture* 15, no. 4 (Winter 1992): 46.

53. Hal Draper, *Berkeley: The New Student Revolt* (New York: Grove Press, 1965), 153.

CHAPTER 12

1. B. F. Skinner, *A Matter of Consequences* (New York: Alfred A. Knopf, 1983), 203.

2. Daniel Bjork. *B. F. Skinner: A Life* (New York: Basic Books, 1993), 178. Bjork is citing "Mistreatment" from November 19, 1986 from the Skinner basement archives.

3. Arthur I. Gates to B. F. Skinner, April 17, 1962, The Papers of Burrhus Frederic Skinner, Harvard University Archives (hereafter, Skinner Papers).

4. Bjork, *B. F. Skinner*, 187.

5. Patrick Suppes, "The Uses of Computers in Education," *Scientific American* 215, no. 3 (September 1966): 207.

6. Suppes, "The Uses of Computers in Education," 208.

7. B. F. Skinner, *The Technology of Teaching* (New York: Meredith Corporation, 1968).

8. Bjork, *B. F. Skinner*, 186–187.

9. Bjork, 186. Bjork is citing a note from the Skinner basement archives titled "Organum" and dated September 19, 1966.

10. Jerome Bruner, *The Process of Education* (New York: Vintage Books, 1960); Bruner, *Toward a Theory of Instruction* (Cambridge, MA: Harvard University Press, 1966).

11. Sidney Pressey, "Teaching Machine (and Learning Theory) Crisis," *Journal of Applied Psychology* 47, no. 1 (February 1963): 5.

12. Pressey, "Teaching Machine (and Learning Theory) Crisis," 2.

13. Pressey, 5.

14. John D. Green, "B. F. Skinner's Technology of Teaching," *Classroom Computer Learning* (February 1984): 24.

15. B. F. Skinner, "Programmed Instruction Revisited," *The Phi Delta Kappan* 68, no. 2 (October 1986): 110.

16. The National Commission on Excellence in Education, *A Nation at Risk: The Imperative for Educational Reform*, April 1963, 7.

17. Gregory Byrne, "U.S. Students Flunk Math, Science," *Science* 243, no. 4892 (February 10, 1989): 729.

18. B. F. Skinner, "Teaching Machines" (Letter to the Editor), *Science* 243, no. 4898 (March 24, 1989): 1535.

19. Christopher Phillips, *New Math: A Political History* (Chicago: University of Chicago Press, 2014), 1.

20. Bruner, *The Process of Education*, 33.

21. B. F. Skinner, "Cognitive Science and Behaviorism," *British Journal of Psychology* 76 (1985): 292.

22. Skinner, "Cognitive Science and Behaviorism," 293.

23. Skinner, 291.

24. B. F. Skinner, "The Shame of American Education," *American Psychologist* (September 1984): 949.

25. Skinner, "Cognitive Science and Behaviorism," 299.

26. B. F. Skinner, *Beyond Freedom and Dignity* (Indianapolis: Hackett Publishing, 1971), 21.

27. Skinner, *Beyond Freedom and Dignity*, 14.

28. Skinner, 59.

29. "B. F. Skinner Says We Can't Afford Freedom," *Time*, September 20, 1971.

30. "Skinner's Utopia: Panacea, or Path to Hell," *Time*, September 20, 1971, 52.

31. Skinner, *Beyond Freedom and Dignity*, 165.

32. "Skinner's Utopia," *Time*, 52.

33. "Skinner's Utopia," 52.

34. Noam Chomsky, "*Verbal Behavior* (Review)" *Language* 35 (1959): 26–58; B. F. Skinner, *Verbal Behavior* (New York: Appleton-Century-Crofts, Inc., 1957).

35. Noam Chomsky, "The Case Against B. F. Skinner," *New York Review of Books* 17 (December 30, 1971), https://www.nybooks.com/articles/1971/12/30/the-case-against-bf-skinner/.

36. Chomsky, "The Case Against B. F. Skinner."

37. Henry L. Roediger III, "What Happened to Behaviorism," *Observer*, March 2004, https://www.psychologicalscience.org/observer/what-happened-to-behaviorism.

38. *A Clockwork Orange*, Box Office Mojo, IMDb, https://www.boxofficemojo.com/release/rl441091585/.

39. Stanley Kubrick, dir., *A Clockwork Orange*, Los Angeles, Warner Bros., 1971.

40. Fred M. Hechinger, "A Liberal Fights Back," *New York Times*, February 13, 1972, D2.

41. Stanley Kubrick, "Now Kubrick Fights Back," *New York Times*, February 27, 1972, D1.

42. Anthony Burgess, "The Clockwork Condition," *New Yorker*, May 28, 2012, https://www.newyorker.com/magazine/2012/06/04/the-clockwork-condition.

CONCLUSION

1. Bill Ferster, *Teaching Machines: Learning from the Intersection of Education and Technology* (Baltimore: Johns Hopkins University Press, 2014), 88.

2. Ludy Benjamin, "A History of Teaching Machines," *American Psychologist* (September 1988): 711.

3. Benjamin, "History of Teaching Machines," 711.

4. Carl Rivers, "Computer Confrontation," *Saturday Review*, April 14, 1973, 48.

5. "Adaptive Learning," Dreambox Learning, https://www.dreambox.com/adaptive-learning, accessed October 14, 2019. Other references to Skinner's work as a foundation for contemporary education technology endeavors include: Nuri Kara and Nese Sevin, "Adaptive Learning Systems: Beyond Teaching Machines," *Contemporary Educational Technology* 23, no. 2 (2013): 108–120; Jason Lodge and Melinda Lewis, "Pigeon Pecks and Mouse Clicks: Putting the Learning Back

into Learning Analytics," presented at ASCILITE 2012, Wellington, New Zealand, November 25–28, 2012.

6. Salman Khan and Michael Noer, "The History of Education," November 1, 2012, produced by *Forbes*, YouTube 11:27, https://youtu.be/LqTwDDTjb6g.

7. W. Lee Garner, *Programed Instruction* (New York: The Center for Applied Research in Education, 1966), 2.

8. Paul Saettler, *The Evolution of American Educational Technology* (Greenwich, CT: Information Age Publishing, [1990] 2003), 303.

9. Joy Lisi Rankin, *A People's History of Computing in the United States* (Cambridge, MA: Harvard University Press, 2018), 3. Unless otherwise noted, all emphasis in quotations is in the original.

10. IDEALS, "Progress Report: 1959–60," Coordinated Science Laboratory, University of Illinois, Urbana-Champaign, Illinois Digital Environment for Access to Learning and Scholarship, 11, http://hdl.handle.net/2142/28161.

11. IDEALS, "Progress Report," 47–48. Lawrence Stolurow, a psychology professor at the University of Illinois and the author of the 1961 book *Teaching by Machine*, developed a competing machine to PLATO called SOCRATES (System for Organizing Content to Review and Teach Educational Subjects)—a historical reference and a bit of a jab, no doubt, as Socrates was Plato's teacher. SOCRATES ran on a film strip-based AutoTutor. The university opted to back further development of PLATO instead of Stolurow's machine, but many of the ideas that guided his teaching machine made their way into the design of PLATO's instructional system. John Gilpin, who briefly worked in B. F. Skinner's teaching machine lab and had gone on to work for Bell Labs, also worked at the University of Illinois, Urbana-Champaign (UIUC), developing instructional programs for the PLATO system. For more on the history of PLATO, see Brian Dear's *The Friendly Orange Glow: The Untold Story of the PLATO System and the Dawn of the Cyberculture* (New York: Pantheon Books, 2017).

12. Nicholas Carr, "How Smartphones Hijack Our Minds," *Wall Street Journal*, October 6, 2017, https://www.wsj.com/articles/how-smartphones-hijack-our-minds-1507307811.

13. Shoshana Zuboff, *The Age of Surveillance Capitalism: The Fight for a Human Future at the New Frontier of Power* (New York: Public Affairs, 2019), 296.

14. Zuboff, *The Age of Surveillance Capitalism*, 216, 224.

15. Zuboff, 20.

16. Zuboff, 179–180.

17. Zuboff, 262.

18. Eugene Galanter, ed., "The Ideal Teacher," *Automatic Teaching: The State of the Art* (New York: Wiley & Sons, 1959), 1.

19. W. W. Charters, "Is There a Field of Educational Engineering?," *Educational Research Bulletin* 24, no. 2 (February 14, 1945): 29–37, 56.

20. Cybernetics, "the scientific study of control and communication in the animal and the machine," was developed around the same time as many of the events in this book. It was another mid-twentieth-century fascination with psycho-technologies and giant counting machines. MIT mathematician Norbert Wiener published his book *Cybernetics: Or Control and Communication in the Animal and the Machine* in 1948. Bell Labs mathematician Claude Shannon published his article "A Mathematical Theory of Communication" that same year. That was also the year that B. F. Skinner published his novel *Walden Two*. W. Ross Ashby, a British cybernetician, published his book *Design for a Brain* in 1952. The following year, B. F. Skinner famously visited his daughter's fourth-grade class and observed, with dismay, all the ways in which it violated his ideas of positive behavioral reinforcement, inspiring him to design his first teaching machine. In 1956, Ashby published *An Introduction to Cybernetics*; Skinner signed a contract with IBM to develop his teaching machine; and Elvis first appeared on TV—which matters for context, for the burgeoning youth culture of postwar America would come to question teaching machines and "programmed instruction" and worry that these might be utterly opposed to the education of free citizens. See Norbert Wiener, *Cybernetics: Or Control and Communication in the Animal and the Machine* (Paris: Hermann & Cie, 1948); Claude Shannon, "A Mathematical Theory of Communication," *Bell System Technical Journal* 27, no. 3 (July 1948): 379–423; W. Ross Ashby, *Design for a Brain* (London: Chapman & Hall, 1952); and Ashby, *An Introduction to Cybernetics* (London: Chapman & Hall, 1956).

21. See, for example, William W. Brickman and Stanley Lehrer's collection *Automation, Education, and Human Values* (New York: School & Society Books, 1966). The book includes articles by anthropologist Margaret Mead, the head of Harvard's Graduate School of Education Francis Keppel, and Vice President Hubert Humphrey.

22. Cited in John L. Rudolph, *Scientists in the Classroom: The Cold War Reconstruction of American Science Education* (New York: Palgrave, 2002), 99. The Woods Hole conference was, of course, where Jerome Bruner and other cognitive scientists began to lay the groundwork for the "New Math."

23. Cited in Rudolph, *Scientists in the Classroom*, 99.

24. James D. Finn, "A New Theory for Instructional Technology," *Audio Visual Communication Review* 8, no. 5 (1960): 87.

25. Finn, "A New Theory for Instructional Technology," 87.

26. That educational engineering would, at the very least, mean teachers had less power and control over teaching and learning should hardly come as a surprise. There are obviously gender dynamics at play, noticeable when someone like B. S. Skinner or Sidney Pressey would say that the largely female teaching profession would not go away entirely but would rather still be in the classroom to take care of students' emotional needs. But there were other developments that underscored the dismissal of teachers. Pressey, for example, began to advocate for "adjunct auto-instruction" in lieu of teaching machines. As the name suggests, these devices really were about self-teaching. And Skinner's friend Fred Keller developed the "personalized system of instruction" (PSI) based on Skinner's theories of behavioral engineering in the classroom. But PSI too downplayed the role of the teacher. Indeed, Keller titled his talk at the 1967 American Psychological Association "Good-bye, Teacher." See Fred Keller, "Good-bye, Teacher . . . ," *Journal of Applied Behavior* 1 (Spring 1968): 79–89.

27. James D. Finn, "Teaching Machines: Auto-Instructional Devices for the Teacher" (1960), in *Extending Education through Technology: Selected Writings by James D. Finn*, ed. Ronald J. McBeath (Greenwich, CT: Information Age Publishing, 2004), 106.

28. James D. Finn, "New Techniques for Teaching in the Sixties" (1961), in McBeath, ed., *Extending Education through Technology*, 210.

29. Daniel Tanner, "The Machine Teacher and the Human Learner," in *Programs, Teachers, and Machines*, ed. Alfred de Grazia and David A. Sohn (New York: Bantam Books, 1964), 308.

30. Tanner, "The Machine Teacher and the Human Learner," 303.

31. Seymour Papert, *Mindstorms* (New York: Basic Books, 1980), 19.

32. Papert, *Mindstorms*, 19.

33. Theodore Roszak, *The Cult of Information: The Folklore of Computers and the True Art of Thinking* (New York: Pantheon Books, 1986), 83.

34. Jacques Ellul, *The Technological Society* (New York: Vintage Books, 1964), xxv.

35. Ellul, *The Technological Society*, 6.

36. Ellul, 347.

37. Ellul, 349.

38. Joseph Weizenbaum, *Computer Power and Human Reason* (New York: W. H. Freeman & Co., 1976), 241.

INDEX

A. B. Dick Company, 39
Abrashkin, Raymond, 156
Addie, Lyle, 54
Age of Surveillance Capitalism, The (Zuboff), 251
Air crib, 86–92, 189–190
Aircrib Corporation, 189
Algebra, 139, 159, 168, 170, 171, 173
American Association for the Advancement of Science, 35
American Council on Education (ACE), 70
American High School Today, The (Conant), 116–119, 121, 123
American Institutes for Research, 146
American Psychological Association (APA), 31, 35, 36, 38, 42, 49, 95, 108, 110, 202
American Summerhill Society, 217
American Voting Machine, 99
Analysis of Behavior, The (Skinner), 203
Anthropoid Ape Research Foundation, 88
Appleton-Century-Crofts Inc., 203
Aristotle, 232
Arithmetic of Computers, The (Crowder), 139
Army Alpha, 36
Arps, George, 53
Artificial intelligence, 247
Atlas Weapons Systems, 256
Automatic Teacher, 38, 44, 45, 55, 56, 57, 59, 60, 64, 79, 231
Automation, 16, 17, 39, 49, 150, 162–163, 230
  of education, 10, 36, 42, 45, 71, 120, 140, 141, 142, 152, 154, 163, 220–221, 249, 258
Ayres, Leonard, 41

Back to Basics movement, 7, 237
Baehne, G. W., 73, 75–77
Banking model of education, Freire's concept of the, 226
Barbrook, Richard, 7
Basic Systems, 204
Bechtel, Les, 94
Behaviorism, 19, 21–24, 25–26, 46, 113, 136, 141, 143, 227, 237, 238, 240, 241, 247, 250. *See also* Operant conditioning; Reinforcement
  and behavioral engineering, 123, 214, 215, 225, 240, 251, 254, 255
  and behavioral surplus, 252
  and behaviorist technology, 15, 23, 232, 238, 243, 244, 254
  and behavior management apps, 30, 252
  and behavior modification, 20, 21
  criticisms of, 89, 216, 227–229, 233–234, 239, 240, 243–244, 262
  and experimentation, 27–28, 253
Behavior of Organisms, The (Skinner), 215
Bell Laboratories, 68, 101, 103–104, 248, 299n20
Benjamin, Ludy, 245–246
Better Business Bureau, 92
Beyond Freedom and Dignity (Skinner), 238–240, 243, 253
Bjork, Daniel, 98, 122, 214, 233

Blyth, John, 15–16, 222–224
Bobbs, Julian, 91
Bobbs-Merrill Publishing Company, 91
Boehm, George A. W., 161–162
Bond, Julian, 223
Book of Knowledge, 185, 187
Boring, Edwin G., 27–30
Briggs, Leslie, 146
Britannica Center for Studies in Learning and Motivation, 178
British Journal of Psychology, 238
Brown v. Board of Education of Topeka, 5, 117, 167
Bruner, Jerome, 233, 237, 256
Buchanan, Cynthia, 176–177
Budden, A. J., 210
Burdorf, Donald, 115
Burgess, Anthony, 241–243
Burner, Eric, 222
Bush, George W., 6
Bush, Vannevar, 67

Califone Corporation, 113, 132, 199, 203. *See also* Rheem Manufacturing Company
Californian ideology, 7
Callahan, Raymond, 43, 160, 230, 255
Calvin, Allen, 167–169, 175–177, 191, 196–197, 248
Cameron, Andy, 7
Carnegie Foundation, 71, 116, 117, 167, 176, 177
Carnegie Mellon University, 155

Carr, Nicholas, 251
"Case Against B. F. Skinner, The" (Chomsky), 239–240
Charters, Werrett Wallace, 255, 257
Child and the Curriculum, The (Dewey), 68
Chomsky, Noam, 239–241
Civil Rights Act of 1964, 186
Civitas Learning, 252
ClassDojo, 252
Classical conditioning, 21
Cleveland Trust Company, 41
Clockwork Orange, A, 241–243
Closer Than We Think (Radebaugh), 156–157
Cognitive science, 232, 236–238, 240, 247, 250
Cole, Luella, 37, 53, 60, 275n73
Coleman, James, 186
"Coleman Report," 185–186
Columbia University, 46, 49, 61, 63, 70, 71, 73, 77, 89, 116
  Bureau of Collegiate Educational Research, 49, 70, 72
  Teachers College, 63
Committee of Ten, 2–5, 8, 10, 66
Committee on Scientific Aids to Learning, 67, 116
Common Sense Book of Baby and Child Care, The (Spock), 89
Comptometer, 98–99, 100, 103, 104
Compulsory Mis-education (Goodman), 218

Computational thinking, 260
Computer-assisted instruction, 12, 15, 235, 246, 247, 249–250, 259
Computer Power and Human Reason (Weizenbaum), 262
Computers in education, 3, 11, 13, 14, 156, 230, 232, 234–235, 236, 237, 246, 249, 259–260
Conant, James Bryant, 68, 116–127, 133, 168, 216, 219
Contemporary Psychology, 27, 29
Coolidge, Calvin, 35
Coons, C. V., 196–202, 204
Cornell, Dudley, 181, 191
Courtis, S. A., 46, 56–57
Crowder, Norman, 139–143, 144, 146, 159, 164, 189, 193, 210, 250. *See also* Western Design
Crowell-Collier Publishing Company, 191
Crozier, William J., 122
Cuban, Larry, 65–66, 119
Cubberly, Ellwood, 44, 49
Cult of Information, The (Roszak), 260
Cybernetics, 7, 14, 256, 258, 299n20

Dalton Adding Machine Company, 40
Danny Dunn and the Homework Machine (Abrashkin and Williams), 156

Davis, I. G., 115, 128–130
Democratic National
   Convention of 1964, 225
Denis, Leila Roosevelt, 88
Dewey, John, 68, 96, 161, 219
Diebold Group, The, 223–224
Display Associates, 89
Donnelly, C. J., 188
Doubleday, 139, 192
Downey, Matthew, 70, 71
Dreambox Learning, 247, 252

Earlham College, 146
Eastman, George, 64, 66
Eastman Kodak, 64, 195
Edison, Thomas, 66, 149, 151
Ed-tech imaginary, 10
Ed-tech teleology, 11
Educational Bureau of Portable
   Typewriters, 62, 64
Educational engineering, 255,
   300n26. *See also* Learning
   engineer
Educational Research
   Association, The, 49
Educational Testing Service
   (ETS), 70
Education and Freedom (Rickover),
   96
*Education and the Cult of
   Efficiency* (Callahan), 43, 160,
   230, 255
Eisenhower, Dwight D., 116
Electrical Test-Corrector Co.,
   The, 75
Eliot, Charles, 66
Ellul, Jacques, 10, 250, 261

Encyclopedia Britannica, 185,
   187, 196
   Encyclopedia Britannica Films
      (EBF), 175–178, 191, 197
Encyclopedias, ownership of,
   186–187
English, Horace, 28–30
Epistemology, 260
Evans, James, 137, 180, 191
Everett, John, 175
*Evolution of American Educational
   Technology, The* (Saettler), 247

Facebook, 252
Fairchild, Sherman, 81
Fascism, 220, 242, 243
Ferster, Bill, 245
Ferster, Charles, 22
Feurzig, Wally, 259
Field Enterprises Educational
   Corporation, 191
Fine, Benjamin, 189
Finn, James D., 257
Flesh, Rudolf, 189
Fogg, B. J., 250
Forbes, 1–5, 11, 13
Ford Foundation, 102, 257
   Ford Fund for the
      Advancement of Education,
      93
Foringer & Company, 105, 107,
   168, 171, 196
Fortune, 158, 160–162
Freedom Schools, 226–227, 229,
   263
Freedom Vote Campaign, 225
Freeman, Frank, 61, 63, 64

Freire, Paolo, 226, 229
Fukuyama, Francis, 11

Galanter, Eugene, 145, 255
Gardner, John, 123
Garner, W. Lee, 247
Gates, Arthur I., 231
Gates, Bill, 1, 2, 12
General Mills, 24
General Motors, 59
Gestalt therapy, 218, 227
GI Bill, 6
Gilbert, Thomas, 101, 103, 147, 248
Gilpin, John, 104, 298n11
Glaser, Robert, 145–147, 157, 180, 189, 190, 193
Goodman, Paul, 217–221, 226–229, 253
Google, 252, 253
Gray, John, 189
Great Depression, 56, 61, 63, 76, 162, 230
Grolier, 180, 181, 183–184, 185, 187, 188–189, 190–191, 197
Gross, Robert, 81
Growing Up Absurd (Goodman), 217, 219

Hagerstown, Maryland, 161
Hamilton College, 30, 112, 146, 222
Harcourt, Brace, and Company, 98–103, 105, 111–112, 192, 232
Harlem, New York, 124–126, 133, 168

Hart, Harold, 217
Harvard Educational Review, 24
Harvard University, 3, 36, 66, 68, 116, 146, 237, 239, 251. *See also* Skinner, B. F.
  Batchelder House, 135
  Department of Psychology, 109
  Graduate School of Education, 32, 81, 93
  Harvard Teachers Association, 67
  "Natural Sciences 114," 136, 203
  pigeon lab, 25, 144, 187
Hastings, Reed, 247
Hawthorne effect, 170
Hechinger, Fred, 118, 242
Heir Conditioner Company, The, 90
Highlander Institute, 222
History of Instructional Technology, A (Saettler), 13
Hively, Wells, 135, 146
Holland, James G., 146
Hollins College, 167, 169, 170, 175–178, 191
Holt, Len, 225
Homme, Lloyd, 27, 93, 135, 144, 146, 147, 180, 184, 190, 193, 248
Horace Mann School, 63
Hughes Aircraft, 105, 192

Indiana University, 37, 181, 187
Individualization, 16, 220, 221, 230. *See also* Personalization

Individualization (cont.)
    of education, 9, 10, 14, 16, 44, 66, 68, 69, 79, 138, 151, 153, 154, 221, 226, 230, 232, 254, 259 (*see also* Personalized learning)
Industrial model of education, 2–4, 5
Industrial Revolution, 4, 8, 60, 149
    of education, 31, 33, 60, 149, 245
Institute for Personnel Workers, 69
Instructional design, 135
Intelligence testing, 6, 36, 69, 89, 124, 254. *See also* Standardized testing
International Business Machines Corporation (IBM), 33, 59, 71–73, 75–85, 93–95, 98–106, 137, 139, 230, 232, 248
    IBM 650, 105
    IBM 805 Test Scoring Machine, 78
    IBM Card Verifier, 83–84
    IBM punch card, 230
International Society for Performance Improvement, 248
Introduction to the Use of Standard Tests (Pressey and Cole), 37
Israel, Matthew, 135

Jetsons, The, 163–164
Jewett, Frank, 67
Jobs, Steve, 12
Johns Hopkins University, 186
Johnson, Reynold B., 73–79, 84
Journal of Applied Psychology, 233
Jovanovich, William, 98–99, 102
Judd, J. Weston, 89–92
Jung, Carl, 22

Keller, Fred, 300n26
Kelly, Kevin, 14
Kelman, Herbert C., 239
Kennedy, John F., 89
Kennedy, Robert F., 89
Kennedy, Ted, 89
Keppel, Francis, 32, 81, 93
Khan, Sal, 1–5, 7, 8, 10, 11, 13, 17, 247
Khan Academy, 2, 3, 4, 8, 262
Khrushchev, Nikita, 97
Klaus, David, 27
Klaw, Spencer, 120, 180
Kubrick, Stanley, 241–242

Ladies Home Journal, 85, 87, 89, 92
Laggards in Our Schools (Ayres), 41
Learning engineer, 155. *See also* Educational engineering
Learning Foundations, 249
Lewis, Walter, 131, 132, 195
Libertarianism, 7, 240
Life Adjustment movement, 7
Lockheed Martin, 81, 99
LOGO programming language, 12, 259, 260
Lubar, Steven, 230
Luddites, 232

Ludovico Technique, 241
Lumsdaine, A. A., 145–147, 157, 181, 188
Luxton, Dean, 107, 108, 109, 111, 112, 113

Macmillan, 64, 214
Mallatratt, Gordon, 203–211
Manhattan Project, 122
Mann, Horace, 2, 4
Marietta Apparatus Company, 40
Markograph, 73–75, 79–80
Marling, Karal Ann, 87
Massachusetts Institute of Technology (MIT), 1, 12, 239, 259, 262
Mattox, Clintis, 168
McCallum, John, 99–104, 112
McCarthyism, 6
McDowell, Malcolm, 241–242
McGraw-Edison, 197
McGraw-Hill, 13, 94, 105, 203
Mechanical education, 39, 66–68, 70, 81, 218, 220, 230
Mentalism, 22, 237
Meredith Publishing Company, 204
Metzner, Robert, 114
Meyer Markle, Susan, 93, 103, 104, 105, 135, 136–139, 143, 144, 146, 177, 248, 284n8
Military, role of in the development of education technology, 113, 143, 144, 146, 192, 256, 257
Miller, H. W., 104
Mindstorms (Papert), 259

Mississippi Freedom Democratic Party, 225
Mississippi Freedom Summer, 222, 225–226
Monroe Calculator Company, 105
Moses, Robert Parris, 222–225, 228
Murphy, Daniel, 169

National Academy of Sciences, 67, 256
National Association for the Advancement of Colored People (NAACP), 124
National Association of College Teachers of Education, 49
National Defense Education Act, 95
National Defense Research Committee, 116
National Education Association (NEA), 55, 146, 147, 247
National Research Council, 67
National School Lunch Act, 6
National Society for Programmed Instruction, 143, 188
National Vocational Education Act, 6
Nation at Risk, A, 235
Neill, A. S., 216–217
Netflix, 247
Newman, Edwin B., 109
New Math, 7, 236–237, 241
New York Stock Exchange, 52
No Child Left Behind, 6

Noer, Michael, 1, 3, 8
Northwestern University, 258
Nudge (Thaler), 251
Nudging techniques, 15,
   251–252. *See also* Operant
   conditioning

Ohio State University, 28, 36, 48,
   49, 50, 53, 54, 55, 245, 255
Operant conditioning, 15, 21,
   30, 87, 214, 228, 233, 239,
   241, 259

Papert, Seymour, 12, 259–260
Pask, Gordon, 199
Patents, 30, 33, 42, 48, 76, 77,
   78, 82, 84, 90, 102, 104, 112,
   133, 188, 195, 196, 206, 207,
   208, 211, 212
Pavlov, Ivan, 21, 22
Pedagogy of the Oppressed (Freire),
   229
Pennsylvania Study, 71
*People's History of Computing in
   the United States, A* (Rankin),
   249
Perkins, Paul, 85
Perlstein, Daniel, 226
Personalization, 2, 70, 254–255.
   See also Individualization
Personalized learning, 3, 9,
   10, 18, 254, 300n26. *See
   also* Individualization, of
   education
Persuasive Technology Lab, 250
Phillips, Christopher, 236

Piaget, Jean, 234
Plessy v. Ferguson, 5
Popular Science, 139, 158, 159
Porter, Douglas, 135, 146
Pressey, Sidney, 9, 12, 36, 61,
   64, 76, 79, 84, 142, 146, 149,
   162, 163, 231, 245–246, 249,
   257, 262, 300n26
   and criticism of teaching
     machine movement, 144–
     145, 147–148, 233–234
   and debates about the
     invention of the teaching
     machine, 28–33
   and work with standardized
     testing, 36–37, 41–42
   and work with Welch
     Manufacturing, 47–59
*Principles of Scientific
   Management, The* (Taylor), 43
Process of Education (Bruner),
   233, 237
Programmed instruction, 18,
   103, 121, 146, 161–162, 168,
   183, 195, 201, 205, 206, 207,
   215, 233, 236, 238, 245, 248,
   283n4, 299n20
   branching, 141, 144, 250
   chaining, 176
   criticisms of, 221–222, 226–229
   development of, at Harvard,
     135, 137–139
   development of, at Hollins
     College, 175–178
   development of, by Norman
     Crowder, 140–143

development of, by TMI, 190, 192–193
experiment during Freedom Summer, 222–223, 225–228
experiment in Roanoke, 168, 170–175
fading, 137, 284n11
frames, 83, 135, 137, 148, 182, 183
intrinsic programming, 141, 142, 250
linear programming, 141, 143
as the origin of computer-assisted instruction, 235, 246–247
as the origin of personalized learning, 14, 15
programs, 111, 135, 138, 148, 192
prompting, 137, 284n11
Programmed Logic for Automated Teaching Operations (PLATO), 249–250, 298n11
Programs, Teachers, and Machines (de Grazi and Sohn), 258
Progressive education, 68, 219, 231–232
Project Pigeon, 25, 85
Prussia, 4, 262
  Prussian model of schooling, 2, 3 (*see also* Industrial model of education)
Psychoanalysis, 22, 237
Psychological Warfare Division, 256
Psycho-technologies, 16, 24, 85, 251. *See also* Behaviorism, and behaviorist technology
Public School Administration (Cubberley), 44

Radebaugh, Arthur, 156–157
Radio Corporation of America (RCA), 105
Ramo, Simon, 150–155, 156–157, 165, 232, 255, 257
Rankin, Joy Lisi, 249–250
Reinforcement, 20, 22, 24, 25, 26, 30–31, 46, 110, 144, 183, 225, 234, 239, 241, 248, 251. *See also* Operant conditioning
Remington Typewriter Company, 39, 62
Rheem Califone Corporation, 113–114, 116, 133, 198, 204, 209. *See also* Rheem Manufacturing Company
Rheem Manufacturing Company, 105–107, 109–110, 112–115, 124, 127–128, 130–133, 168, 184, 191, 195, 231–232
  "Didak 101," 110, 198, 202
  "Didak 501," 111, 112, 133, 197, 198, 200, 202, 203, 204, 206, 207, 208, 209, 210
  "Didak 701," 198
Rickover, Admiral Hyman, 96, 216
Rivkin, Donald, 106, 110, 112, 130–132, 133, 199, 212

Roanoke, Virginia, 159, 196, 236, 248
  Roanoke Experiment, 167, 170, 173
Robson, Hal, 105–106
Roosevelt, Franklin D., 122
Roosevelt, Theodore, 88
Roszak, Theodore, 260
Royal Typewriter, 62
Rushton, Edward, 168–175
Rutherford, Alexandra, 23, 87

Saettler, Paul, 13–14, 247
Saltzman, Irving, 135
Savio, Mario, 229
Schedules of Reinforcement (Skinner and Ferster), 22
School and Society (journal), 31, 42, 45, 59, 60, 62, 146
Science (journal), 30, 109, 120, 146, 156, 235
Science and Human Behavior (Skinner), 113, 136, 214
Science News Letter (journal), 27, 31
Scientific management, 43–44
Scopes Monkey Trial, 6
Segregation, 117, 167, 186, 288n3
Seligman, Daniel, 160–161
Shady Hill School, 19
Shriver, Eunice K., 89
Silicon Valley, 1, 7, 15, 249, 251, 254
Simon, Herbert, 155
Skinner, B. F., 12, 13, 15, 18, 76, 120, 146, 147, 156, 158, 162–163, 167, 168–170, 177, 183, 184, 185, 188, 189, 190, 191, 192, 222, 231–238, 257, 262
  and conflict with Conant, 121–127
  and conflict with Crowder, 141–144
  criticism of, 213–216, 218, 239–244
  and debate about invention of teaching machine, 28–33
  and development of first teaching machine, 19–27
  and development of the air crib, 85–92
  influence on contemporary education technology, 245–248, 251, 253–254
  Skinner Box, 24, 87, 263
  and teaching machine group at Harvard, 135–137, 180
  theory of behaviorism, 22–23
  and work with Harcourt Brace, 98–105
  and work with IBM, 80–84, 93–95, 98–106
  and work with pigeons, 24, 25, 31, 144, 187, 234, 239, 253
  and work with Rheem Manufacturing, 107–116, 127–133, 195
Skinner, Halcyon, 30
Smith & Corona, 62
Solomon, Cynthia, 259
Soviet Union, 6, 17, 95–97, 237
Spock, Benjamin, 89

Sputnik, 6, 17, 95–97, 116, 119, 123, 174, 235, 236, 237, 257
Standardization, 2, 3, 16, 43, 66, 67, 132, 216, 219, 220, 259
Standardized testing, 6, 36, 37, 38, 42, 44, 45, 57, 69, 70, 71, 72, 73, 137, 170, 173, 174, 236. *See also* Intelligence testing
Standard Oil Company, 107
Stanford-Binet intelligence test, 89
Stanford University, 44, 49, 232, 250
Star Trek, 11
Stevens, Stanley Smith, 109
Stock market crash, 52
Strong Vocational Interest Blank, 72
Student Nonviolent Coordinating Committee (SNCC), 222–224, 226, 227, 228, 294n30
Sullivan, Maurice, 176–177
Summerhill: A Radical Approach to Child Rearing (Neill), 216–217
Suppes, Patrick, 232, 247
Surveillance capitalism, 252, 254
Sweeney, J. L., 63, 64
Systems thinking, 258
 in education, 255–256, 260

Tanner, Daniel, 258
Taylor, Frederick Winslow, 44, 160
Teachers and Machines (Cuban), 65
TEaching MAChines (TEMAC), 175
Teaching Machines and Programmed Instruction (Glaser and Lumsdaine), 145, 157, 181, 189
Teaching Machines Inc. (TMI), 180–184, 187–191, 193, 224, 248
 Min/Max, 181, 182, 185, 187, 188, 189, 191, 193
 Wyckoff Film Tutor, 188–189
Technique, Ellul's concept of, 10, 261
Technological Society, The (Ellul), 261
Technology of Teaching, The (Skinner), 233
TED Talks, 1, 8
Teleology of ed tech, 11
Terman, Lewis, 89
Thaler, Richard, 251
Thomson Ramo Wooldridge Inc., 150
Thoreau, Henry David, 213–214
Thorndike, Edward, 46, 70, 89
 laws of learning, 46
Tinkering Toward Utopia (Cuban and Tyack), 119
TMI Institute, 190
Tosti, Donald, 190
Tougaloo College, 223–224
Toward a Theory of Education (Bruner), 233
Truman, Harry, 122
Tutoring, 1, 12, 138–139, 142, 223, 232, 255

TutorTexts, 139–141, 193
Tyack, David, 8, 119

Underwood Typewriter
    Company, 40, 62
Union Thermo-Electric
    Corporation, 99, 102
University of Buffalo, 93
University of California,
    Berkeley, 229–230
University of California, Los
    Angeles (UCLA), 146
University of Chicago, 61
University of Georgia, 101
University of Illinois Urbana-
    Champaign, 249
University of Michigan, 144
University of Pennsylvania, 144
University of Pittsburgh, 24, 27,
    93, 180
University of Utah, 146
Univox Institute, 180
Urban, Wayne, 95
US Air Force, 256
  Office of Scientific Research,
    144
US Army Psychological Corps,
    36, 70
U.S. Industries, 143, 210, 211
US Office of Education, 125,
    159, 176
US Patent Office, 196

Varela, Mary, 227–228
Verbal Behavior (Skinner), 113,
    239
Vietnam War, 238

Virginia Department of
    Education, 172

Walden (Thoreau), 213
Walden Two (Skinner), 214–216,
    228, 238, 253
Waller, Theodore, 183–184
Watson, A. G., 40–41
Watson, John B., 22
Watson, Thomas, Jr., 105
Watson, Thomas, Sr., 71, 78, 99
Weizenbaum, Joseph, 262
Welch, Medard, 48, 49, 55, 59
Welch, Richard, 51, 53, 57
Welch, William, 47, 50
Western Design, 105, 143. *See
    also* U.S. Industries
  AutoTutor, 164, 189, 193,
    210, 211 (*see also* Crowder,
    Norman)
Westinghouse Electric
    Corporation, 190
Why Johnny Can't Read (Flesh),
    189
Wiener, Norbert, 258
Wiggam, Albert Edward, 62
Williams, Jay, 156
W. M. Welch Manufacturing,
    47–49, 51–53, 55, 57–58, 64,
    84, 230, 248
Wood, Ben D., 49, 61–64, 66–72,
    77–79, 116, 148, 219, 230,
    254
Wood-Freeman report, 63, 64
Woods Hole, Massachusetts, 237,
    256, 299
World Book Encyclopedia, 191

World War I, 6, 36, 41, 70, 256
World War II, 24, 85, 213, 256
Wyckoff, Ben, 181, 187–188,
 190, 193, 224, 248

Yale University, 49
Yerkes, Robert, 36
Youngdale, George, Jr., 94–95,
 104

Zenner, R. E., 99, 102
Zuboff, Shoshana, 251–254